Selected Topics in Preventive Cardiology

ETTORE MAJORANA INTERNATIONAL SCIENCE SERIES

Series Editor:
Antonino Zichichi
European Physical Society
Geneva, Switzerland

(LIFE SCIENCES)

A Continuation Order Plan is available for this series. A continuation order will bring delivery of
each new volume immediately upon publication. Volumes are billed only upon actual shipment.
For further information please contact the publisher.

Selected Topics in Preventive Cardiology

Edited by

Angelo Raineri

University of Palermo Medical Center
Palermo, Italy

and

Jan J. Kellermann

The Chaim Sheba Medical Center
Tel-Hashomer, Israel

Plenum Press • New York and London

Library of Congress Cataloging in Publication Data

Main entry under title:

Selected topics in preventive cardiology.

(Ettore Majorana international science series. Life sciences; v. 12)
"Proceedings of the second course of the International School of Cardiology, held
May 10–16, 1982, in Erice, Sicily, Italy"—Verso t.p.
Includes bibliographical references and index.
1. Coronary heart diseases—Prevention—Congresses. 2. Heart—Diseases—Pre-
vention—Congresses. I. Raineri, A. II. Kellermann, Jan J. III. International School of
Cardiology. IV. Series: Ettore Majorana international science series. Life science; 12.
[DNLM: 1. Coronary disease—Prevention and Control—Congresses. 2. Myocardial
diseases—Prevention and control—Congresses. 3. Cardiology—Congresses. W1
ET712M v. 12/WG 300 I635 1982s]
RC685.C6S45 1983   616.1′2305   83-8989
ISBN-13:978-1-4613-3738-6   e-ISBN-13:978-1-4613-3736-2
DOI: 10.1007/978-1-4613-3736-2

Proceedings of the Second Course of the International School of Cardiology,
held May 10–16, 1982, in Erice, Sicily, Italy

©1983 Plenum Press, New York
Softcover reprint of the hardcover 1st edition 1983

A Division of Plenum Publishing Corporation

233 Spring Street, New York, N.Y. 10013

PREFACE

The aim of the 2nd Course of the International School of Cardiology at "Ettore Majorana" was the discussion, scientific analysis, and critical appraisal of primary and secondary prevention in cardiology, especially concerning the coronary artery disease.

The predictive value of different risk factors, the problem of heredity, sociological and psychological perspectives, the impact of antiarrhythmic therapy, and the prevention of sudden death were only some of the many topics discussed.

Others, such as physical activity in primary and secondary prevention, the role of coronary bypass surgery in decreasing the mortality rate, the effects of drugs and comprehensive intervention programmes, as well as a critical appraisal of therapy in cardiac failure were given special attention.

In organizing this second Course, it was again a privilege and a challenge for the programme directors to conduct a scientific event which, as was demonstrated at the first course, represented an original and new approach to teaching. The illustrious international faculty, composed of well known scientists and experts in their fields, has contributed to the high academic level and quality of the lectures and the following discussions.

In our opinion, this successful undertaking would not have achieved its purpose and originality without the beautiful surrounding scenery and tranquility of Erice.

We should like to express our gratitude to the director of the Centre, Prof. Zichichi, and the Secretary of "Ettore Majorana," Dr. Gabriele, for their outstanding cooperation. The editors hope that a glimpse of the spirit of this second Course will be reflected in the volume.

Jan J. Kellermann
Angelo Raineri

Directors of the Course on
Prevention in Cardiology

CONTENTS

INTRODUCTION

The International School of Cardiology is engaged, as the other Schools in this Centre, in promoting scientific and cultural meetings.

In this direction we began our activity in 1979 with the Course on "Functional Evaluation and Rehabilitation in Cardiology."

Following this aim we are here to discuss the "Prevention in Cardiology."

On this occasion too Prof. Jan Kellermann, who I would like to thank deeply, gives us all his experience as a scientist and as a very keen organizer of cultural initiatives to offer this second Course every prerogative of success.

In the past culture was an estimated quality, a type of conscious ideal of human perfection; now it is considered as the social contest in which we are born and live.

Culture, in a sense would no longer be what we are trying to reach or to be, but the historic and sociological reality in which we are living and growing.

Stimulated by natural desire to search, discover and know which is typical of human beings, the modern cardiologist can use instruments of high technological significance so that the investigation, observation and interpretation of events can happen with the best penetrating capability.

Therefore it is necessary to stress that if as researchers we are prone to consider the advantages technology has given to scientific progress, on the other hand it is not possible to neglect that in general the transfer of technological development on the contest of social life has determined a situation in which the way of thinking has been modified, creating an upheaval in the habits of life. Thus nowadays, perhaps for this, we face a deep change in the spectrum of pathologies responsible for the causes of death, in which there is the prevalence of the chronic-degenerative component.

1

The profound changes of pathology during the last decade impose a radical change of purposes. The efforts, that up to now have been predominantly addressed toward a more important widening of our knowledge of the aspects of clinical diagnosis and therapy, have to be moved back towards the prevention.

In the Course we have got epidemiologists, cardiologists and cardiac surgeons together with the aim of giving a multidisciplinary order to a topic that is a strategic choice in confronting a problem of a great practical importance.

Firstly I would like to thank all the lecturers for showing a great sensitivity in accepting our invitation.

I would like to wish all the participants proficous work, in deepening their knowledge.

Angelo Raineri

Director of International
School of Cardiology
"Ettore Majorana Center"

HEREDITY, ATHEROSCLEROSIS AND

CORONARY HEART DISEASE

Frederick H. Epstein

Institute of Social and Preventive Medicine
University of Zürich
CH 8006 Zürich

The familial-genetic aspects of coronary heart disease (CHD) are something of a stepchild in atherosclerosis research for, probably, two major reasons: (1) family studies are tedious and difficult, (2) professional geneticists who are needed as collaborators in this kind of work, sometimes tend to shy away from complex, polygenic situations, with the added problem of having to disentangle the influence of heredity from environmental factors. Yet, there is a trend toward including families in clinical-epidemiological investigations and genetic models to cope with the difficulties mentioned are beginning to be developed[1,2]. The statement has been made for many years that "the family history is important" with regard to susceptibility toward CHD but there is still no entirely satisfactory answer to the question how important it is quantitatively as compared with other risk factors and how importance relates to the age at which the disease presents[3-5].

In practice, it is asked to what extent coronary heart disease aggregates in families. This review will deal mostly with this question, rather than the equally important but relatively better studied matter of familial resemblance in CHD risk factors. However, even in this latter instance, much more is known about familial transmission in the top range of risk factor levels (e.g. the hyperlipidaemias) than in the mild and moderate range where most cases of CHD occur. The term "familial" rather than "genetic" has been used deliberately and consistently because families share their living habits as well as their genes and it is notoriously difficult to separate the two (the "nature-nurture" problem).

Important though it is to learn more about the risk carried by oneself if a close relative is affected with CHD, the statement not

3

infrequently heard that hereditary predisposition is the most important CHD risk factor is to be deplored. This is an unfortunate half-
truth which detracts attention from the paramount influence of the
environment on the causes and prevention of the disease. Of course,
in a constant environment, biological variation is determined entirely by genetic predispositions. In so-called developed and more
and more in developing countries, the majority of people in the
population have adopted patterns of daily living which promote
atherosclerosis and premature coronary heart disease. In such an
all-pervading, detrimental environment, genetic factors must of
necessity exert a decisive influence. On the other hand, there is
every reason to believe that the big differences in the frequency
of coronary heart disease between populations differing in living
habits are largely determined by the environment, as are the marked
secular changes which have occurred during the last decades, both
upward and downward, in many countries.

Familial Clustering of Coronary Heart Disease

     The earlier literature on this subject has been reviewed[2-6]
and attention will be confined to more recent data. A Finnish Study
is particularly instructive[7] (Table 1). The probands were all men
who developed CHD before the age of 56. In one of the analyses
shown, it is apparent that a brother of a patient with myocardial
infarction prior to age 56 has a very high risk of dying of the
disease himself by age 60, being around 7 times higher than the
risk of a brother without such a "family history"; sisters also
carry a higher risk, as do brothers of male probands with angina
pectoris. In another analysis of the same series[8], the risk of a
parent dying before age 70 is seen to be dependent on the age of
occurrence of MI in the son, rising steeply with decreasing age
(Table 2). Family studies amongst Japanese living in Hawaii also
suggest a decisive influence of the age at which CHD presents in the
proband on the risk in his relatives[9]; the risk of developing CHD
in siblings of an index case is doubled in early-onset (age 52 or
less) cases while no correspondingly increased risk is associated
with later-onset cases. Twin studies are especially valuable in
differentiating environmental from genetic transmission, even though
they do not eliminate the problem. The concordance rate for CHD is
higher in monozygotic than in diziygotic twins of the same sex, indicating that a genetic factor is involved[10]; when the twins are of
unlike sex, concordance is very high if the index case is female[10].
The latter finding again points toward a strong genetic element
because a women who breaks through the barrier of her relative protection against CHD must, presumably, have a strong hereditary predisposition.

Table 1. Cumulative probability of fatal coronary heart disease (%)
          before age 60

	Probands with:			Reference Persons
	Fatal myocardial infarction (MI)	Non-fatal MI	Angina Pectoris	
Brothers	34.5%	28.9%	15.4%	4.4%
Sisters	8.2%	3.5%	-	-

After Rissanen, Brit. Heart J. 42:294, 1979

Table 2. Parents of probands: Cumulative probability of dying (%)
          from coronary heart disease by age 70

	FATHER		MOTHER	
Age of Proband at first myocardial infarction	South Finland	East Finland	South Finland	East Finland
≤ 45	43.3	52.3	35.8	24.9
46-50	29.5	35.9	14.5	11.3
51-55	12.1	20.2	6.7	13.2
Reference Group	9.4	7.6	10.0	7.4

After Rissanen, Am. J. Cardiol. July 1979

Familial Resemblance in Risk Factors

Accepting the fact that relatives of CHD patients carry themselves a higher risk of CHD, at least if the disease presents relatively early in life in the proband, the question arises whether
this increased risk is due to the further fact that elevated risk
factors are more common both in the probands and their relatives
because of familial resemblance in risk factors. How strong is this
resemblance? There is a rule that rare, single genes show strong
familial aggregations while common, polygenic traits are associated
with weaker resemblance amongst relatives. The familial hyperlipidaemias illustrate this generalization. If risk factors like serum
cholesterol or blood pressure are viewed as continuous variables,
as they should be, there is familial resemblance along the whole
range but the correlation coefficients between siblings or parents
and children, though significant, are not high and of the order of,
perhaps, 0.15-0.25[11-13].. Against this background, it may now be
asked to what extent the presence of a major risk factor in a CHD

proband affects CHD risk in a close relative and what proportion of
this risk may be mediated by one or more of these risk factors.

The Role of Risk Factors in Mediating Familial Coronary Heart Disease Risk

The presence of hyperlipoproteinaemia in an index patient with
CHD increases CHD risk in a first degree relative[14] (Table 3). The
risk is relatively greater in younger relatives and younger index
patients transmit higher risk even with "normal" lipid levels, remembering that "normal" in this analysis is really a misnomer, meaning merely that no Type II or IV hyperlipoproteinaemia is present.
More recently, the increased CHD risk of Type II relatives of probands with Type II hyperlipoproteinaemia has again been well documented[15].

Another study in Finland[16] (Table 4) shows not only that hypertension and hypercholesterolaemia are much more common in a CHD proband with a CHD family history but that the same traits are also a
great deal more frequent in his siblings (no such relationship exists
for hypertriglyceridaemia). On the basis of these data, the authors
believe that the two risk factors must largely account for familial
clustering of CHD. Data from the Tromsø Study, however, do not bear
out this conclusion since risk factors do not differ amongst persons
with and without a family history of CHD[17] (Table 5).

Before proceeding, attention should be drawn to studies which
address the question whether risk factor levels in childhood bear a
relationship to the presence or absence of a family history of CHD
in their adult relatives. When schoolchildren are divided into
those whose serum cholesterol levels are in the highest and lowest
5% of the distribution and the rest with levels in between, myocardial
infarction is more common in the male and female relatives of the
children in the top range[18]. However, in another study, taking as
point of departure schoolchildren with and without a parental history
of CHD, average serum cholesterol levels did not differ but triglycerides were higher and HDL-cholesterol levels lower in children
of CHD patients than controls[19]. Yet, as a rule, hypercholesterolaemia is common amongst children of young patients with CHD[20].

Why does Coronary Heart Disease aggregate in Families?

It may be tentatively concluded, up to this point, that at
least part of the reason why CHD clusters in families lies in the
fact that relatives of CHD probands are more often hyperlipidaemic
or hypertensive than people in the population at large. How much
of the clustering, however, remains unexplained? A number of years
ago, it was suggested, using a model based on existing epidemi-

Table 3.   Increased Risk of Death to first-degree Relatives of
Patients with Coronary Heart Disease

Index Patients	First Degree Relatives	
	All Relatives	"Younger Relatives"
Whole Group	x 1.5*	x 2.5**
"Younger", with normal lipids	x 1.9*	x 2.4**
Type II Hyperlipoproteinaemia	x 2.5**	x 4.9**
Type IV "	x 1.9	x 4.4**
Spouses with normal lipids	x 0.9	x 0.9

* $p < 0.05 > 0.01$ After Patterson and Slack, 1972

** $p < 0.01 > 0.001$

"Younger" means $<$ 55 years for men and $<$ 65 years for women.

Table 4.   Frequency of Hypertension and Hyperlipidaemia in case
sibs according to parental history of Coronary Heart
Disease (CHD)

	HYPERTENSION %			HYPERCHOLESTEROLAEMIA %			HYPERTRIGLYZERIDAEMIA** %		
	Proband	Brothers	Sisters	Proband	Brothers	Sisters	Proband	Brothers	Sisters
Mother died of CHD before age 70 (14 families)	70.0	37.5	43.8	60.0	45.8	21.9	10.0	20.1	18.8
Father died of CHD before age 70 (38 families)	56.2	29.9	26.9	26.1	17.9	22.4	30.4	19.4	17.1
Neither parent died of CHD before age 70 (108 families)	33.3	16.4	23.6	30.6	21.4	13.5	26.4	15.7	22.5

After Rissanen and Nikkila, Brit. Heart J. 42:373, 1979

*   age-adjusted serum cholesterol $\geqslant$ 8.8 mmol/l (men and women)

**   age-adjusted serum triglyzerides $\geqslant$ 2.3 mmol/l (men); $\geqslant$1,7 mmol/l (women)

ological data, that this fraction might be substantial[4]. Quite
recently, in a retrospective study, it was concluded that the family
history discriminated much more strongly between CHD patients and
matched controls than any other of a large series of risk factors
measured[21]. In the first prospective study where the family history
of CHD was included in multivariate analysis comprising all the other
major risk factors, it emerged as an independent predictor as far
as CHD in the father though not the mother was concerned[22].

Table 5. Age-adjusted mean values of Coronary Heart Disease risk
          factors in subjects with and without myocardial infarction
          (MI) among their first degree relatives

The Finnmark Study (Tromsø)

	MALES		FEMALES	
	MI	No MI	MI	No MI
Cholesterol (mg/dl)	275.0	265.7	264.0	260.5
Triglyzerides (mmol/l)	2.20	2.15	1.56	1.58
Syst. B.P. Pressure	138.1	134.8	122.9	124.6
Diast. B.P. Pressure	81.8	80.7	75.9	76.1
Rel. Body Weight	2.4	2.4	2.3	2.3
Daily No. Cigarettes/ person	8.3	8.8	5.6	4.7

After Thelle and Førde, Am. J. Epidemiol. 110:708, 1979

N.B.: In the "Tromsø Study", the relative risk for MI in first degree relatives of
       MI patients ranged from 5.5 to 12.8, depending on the sex of the patient.

The latest contribution to the question raised comes from the
Framingham Study[22]. It is based on a 26-year follow-up period and
uses, for the first time, clinical-epidemiological diagnoses of CHD
as evidence for a positive family history rather than a hearsay report from a family member with all its sources of unreliability.
For pairs of brothers who were both examined, the occurrence of CHD
in an older brother was a significant predictor of CHD in his younger
brother, even when age, blood pressure, serum cholesterol, smoking
and relative weight were taken into account by multivariate analysis.
The excessive risk carried by a positive history is 1½ to 2-fold but
the analysis does not permit a statement how much of this excess is
referable to the family history alone. The proportion cannot be
large in absolute terms and it must be remembered that most of the
CHD probands in the Framingham cohort are middle-aged or older.

Summary and Conclusion

Viewing the evidence presented in broad perspective, a family
history of CHD in a close relative emerges as a CHD risk factor,
partly independent of the concurrent risks carried by others like
elevated levels of blood pressure, serum cholesterol and smoking.
The risk inherent in a positive family history appears to be largely
but not entirely limited to the familial occurrence of early-onset
CHD. It is not known what risk factors other than those mentioned
might contribute to the familial clustering of CHD, remembering
that familial resemblance in risk factors might be determined by
genetic factors, shared environmental influences or both; they range
from mechanisms related to thrombosis-haemostasis to the risks

associated with various types of psychosocial stress. Genetic predispositions mediated through the immune system, reflected by HLA-
patterns, or indicated by markers like the blood groups must also
be kept in mind. It is likely that further systematic studies of
CHD and risk factors in families selected from the population or
special target groups will bring new insights into the mechanisms
of atherosclerosis and its clinical consequences.

REFERENCES

1. W.J. Schull and K.W. Weiss, Genetic epidemiology: four
    strategies, Epidemiologic Reviews 2:1 (1974).
2. C.F. Sing and M. Skolnick (eds.), Genetic Analysis of Common
    Diseases: Applications to Predictive Factors in Coronary
    Disease, Alan R. Liss, Inc., New York (1979).
3. F.H. Epstein, Hereditary aspects of coronary heart disease,
    Am.Heart J. 67:445 (1964).
4. F.H. Epstein, Risk factors in coronary heart disease - environ-
    mental and hereditary influences, Isr.J.Med.Sci. 3:594 (1967).
5. F.H. Epstein, Genetics of ischaemic heart disease, Postgrad.
    Med.J. 52:477 (1976).
6. Task Force on Genetic Factors in Arteriosclerotic Diseases,
    DHEW Publication No. (NIH) 76-922, Washington, DC, US GPO (1976).
7. A.M. Rissanen, Familial aggregation of coronary heart disease
    in a high incidence area (North Karelia, Finland), Brit.Heart.J.
    42:294 (1979).
8. A.M. Rissanen, Familial occurrence of coronary heart disease:
    effect of age at diagnosis, Am.J.Cardiol. 44:60 (1979).
9. R.L. Phillips, A.M. Lilienfeld, E.L. Diamond and A. Kagan,
    Frequency of coronary heart disease and cerebrovascular accidents
    in parents and sons of coronary heart disease index cases and
    controls, Am.J.Epidemiol. 100:87 (1974).
10. J. Slack, Genetic differences in liability to atherosclerotic
    heart disease, J.Roy.Coll.Phycns.London 8:115 (1974).
11. B.C. Johnson, F.H. Epstein, and M.O. Kjelsberg, Distributions
    and familial studies of blood pressure and serum cholesterol
    levels in a total community - Tecumseh, Michigan, J.Chronic
    Dis. 18:147 (1965).
12. R.J. Havlik, R.J. Garrison, M. Feinleib, W.B. Kannel, W.P.
    Castelli, and P.M. McNamara, Blood pressure aggregations in
    families, Am.J.Epidemiol. 110:304 (1979).
13. M. Feinleib, R.J. Garrison, N.O. Borhani, R.H. Rosenman, and
    J. Christian, Studies of hypertension in twins, in: "Epidemiology
    and Control of Hypertension," O. Paul, ed., Stratton, New York
    and London, p.3 (1975).
14. D. Patterson and J. Slack, Lipid abnormalities in male and
    female survivors of myocardial infarction and their first degree
    relatives, Lancet 1:393 (1972).

15. N.J. Stone, R.I. Levy, D.S. Fredrickson, and J. Verter, Coronary artery disease in 166 kindred with familial type hyperlipoproteinemia, Circulation 49:476 (1974).

16. A. Rissanen and E. Nikkila, Aggregation of coronary risk factors in families of men with fatal or non-fatal coronary heart disease, Brit.Heart J. 42:373 (1979).

17. D.S. Thelle and O.H. Førde, The Cardiovascular Study in Finnmark County: Coronary risk factors and the occurrence of myocardial infarction in first-degree relatives and in subjects of different ethnic origin, Am.J.Epidemiol. 110:708 (1979).

18. H.G. Schrott, W.R. Clarke, D.A. Wiebe, W.E. Connor, and R.M. Lauer, Increased coronary mortality in relatives of hypercholesterolemic school children: The Muscatine Study, Circulation 59:320 (1979).

19. D. Pometta, H. Micheli, L. Raymond, I. Oberhaensli, and A. Suenram, Decreased HDL cholesterol in prepubertal and pubertal children of CHD parents, Atherosclerosis 36:101 (1980).

20. C.J. Glueck, P.M. Laskarzewski, D.C. Rao, and J.A. Morrison, Familial aggregation of coronary risk factors, in: "Complications of Coronary Heart Disease," W.E. Connor and D. Bristow, eds., Lippincott Co., Philadelphia (1982).

21. J.J. Nora, R.H. Lortscher, R.D. Spangler, A.H. Nora, and W.J. Kimberling, Genetic-epidemiologic study of early-onset ischemic heart disease, Circulation 61:503 (1980).

22. C.B. Snowden, P.M. McNamara, R.J. Garrison, M. Feinleib, W.B. Kannel, and F.H. Epstein, Predicting coronary heart disease in siblings - a multivariate assessment, The Framingham Study, Am.J.Epidemiol. 115:217 (1982).

PRIMARY PREVENTION OF CORONARY HEART DISEASE

M. Kornitzer

Laboratory of Epidemiology and Social Medicine
Free University of Brussels
Campus Erasme - CP 590, Route de Lennik 808
1070 Brussels, Belgium

Atherosclerosis or better said atherothrombosis is the pathological basis for coronary heart disease (C.H.D.). Autopsy studies
have shown a correlation between the prevalence of atherothrombotic
complicated lesions and coronary mortality at the population level.
On the other hand correlations between complicated lesions and the
major coronary risk factors like hypercholesterolemia, high blood
pressure or smoking have also been observed.

Fatty streaks appear in early life even during the first decade
and develop during the second and third decades whilst fibrotic and
complicated lesions appear at a subclinical level. Starting during
the third and increasing during the following decades the clinical
manifestations of atherosclerotic disease appear: angina, myocardial
infarction (M.I.), sudden death (S.D.), cerebrovascular incidents
(T.I.A., stroke), renal insufficiency, peripheral arterites: it is
the proverbial tip of the iceberg.

Ideal, and for the moment utopic, prevention would rest on the
total disappearance of atherosclerosis, at least the fibrotic and
complicated lesions, at the population level. A less ambitious
goal would be the persistence of atherosclerosis at a subclinical
level during the first six or seven decades of mens' life delaying,
appearance of clinical manifestations like angina, myocardial infarction or sudden death.

At each moment of the natural history of C.H.D. the possibility
of preventive action has been and is still investigated. In the
Whitehall Study the major coronary risk factors, hypercholesterolemia,

high blood pressure and cigarette smoking were still powerful predictors of coronary mortality in a group of subjects with angina
pectoris and/or minor E.C.G. modifications[1], but no randomized
trials are available to prove the preventability of M.I. and sudden
death in these subjects. In angina pectoris the surgical approach
for angiographically defined subsets of patients can not be dismissed as has been shown by several randomized trials. The prevention of M.I. in unstable angina rests maybe on the administration
of heparine, intravenously, as has been shown in a recent randomized
trial.[2] Lately the preventability of M.I. based on the very early
administration of I.V. Betablockers has been and is still investigated. Last but not least, the modification of the natural history
of M.I. has been extensively studied by means of randomized trials
using betablockers and drugs acting on platelet aggregation.

Causes for coronary heart disease are multiple and gradual,
neither necessary nor sufficient. The assumption of causal relationship should possibly rely on randomized trials: they are technically
difficult, expensive, time consuming and the ones published all open
to criticism. For most of the numerous so-called coronary risk
factors (Table 1) the causal relationship with C.H.D. rests on eight
criteria for causal inference:

1. The power of association: relative risk should be elevated.
2. A dose/effect relationship.
3. A temporal relationship: the factor should precede the
   event.
4. Same observations should have been made in several epidemiological studies.
5. The factor should be independently associated with the
   event: multivariate analysis will solve this problem.
6. The factor should bear a predictive power: its level should
   predict the incidence of C.H.D. in other population studies.

Table 1.   Coronary Risk Factors

Age	Blood viscosity
Sex	Heart rate
Serum cholesterol	Sedentarity
Blood pressure	Water hardness (/-/)
Cigarette smoking	Oral contraceptives
HDL-cholesterol (/-/)	Urbanisation
Triglycerides	Type "A" behavior pattern
Glucose intolerance	Neuroticism
Diabetes	Stress
Obesity	Genetic factors
Nutrition (Fibers /-/;	
Linoleic Acid /-/)	

Table 2.   Criteria for Inference of Causality

	Serum Cholesterol	Triglycerides
1. Power of association	+	+
2. Dose/effect relationship	+	+
3. Temporal relationship (Factor → Event)	+	+
4. Same observations in several studies	+	±
5. Association is independent from other factors	+	−
6. Predictive power	+	−
7. Same observations in animal studies	+	−
8. There is a logical pathogenetic mechanism	+	−

    7.   Same observations in animal studies: experimental studies should have come to the same conclusions.

    8.   There should be a pathogenetic mechanism.

Let us have a look at those 8 criteria for serum cholesterol and triglycerides (Table 2). For serum cholesterol all eight of them are fulfilled. For serum triglycerides, 3 out of 8 are fulfilled: whereas the same observation has been made in some epidemiological prospective studies this has not been the case in all of them. Moreover, the association of triglycerides with C.H.D. is not independent from other factors nor is there a predictive power. No observations have been published in animal studies and no pathogenetic mechanism is at hand.

For high blood pressure 7 out of 8 criteria are positive while some discussions are still going on the pathogenetic mechanism (Table 3).

Table 3.   Criteria for Inference of Causality

	High blood pressure
1. Power of association	+
2. Dose/effect relationship	+
3. Temporal relationship (Factor → Event)	+
4. Same observations in several studies	+
5. Association is independent from other factors	+
6. Predictive power	+
7. Same observations in animal studies	+
8. There is a logical pathogenetic mechanism	±

For cigarette smoking things are less clear as one can see on Table 4. Not all epidemiological prospective studies have shown an association with C.H.D.: Keys et al. did not find an independent relationship in the 10 year follow-up of the southern European cohorts of the Seven Countries Studies[3], hence there is not necessarily a predictive power. In animal studies, the prerequisite for an influence of nicotine is the preliminary administration of an atherogenetic diet. Last but not least the pathogenetic mechanism is still under heavy discussion: carbon monoxide or nicotine or both of them.

In summary we can say that for the three major coronary risk factors, high serum cholesterol, high blood pressure and cigarette smoking causal relation with C.H.D. although not absolutely proven is highly probable.

Let us try to answer two other questions:
1. How? Do we master the techniques for permanent modification of the coronary risk profile?
2. When? When should preventive advice be given?

How to modify the coronary risk profile? We are facing two strategies which are of course not mutually exclusive: the individual approach like in the Multiple Risk Factor Intervention Trial[4] or the Oslo Study[5] focusing on high risk subjects and on the other hand the mass intervention approach like the North-Carelia Project[6]. The latter approach is compatible with slow and gradual risk factor modification at the population level but has some major socio-economic implications, in terms of mass modification of nutritional habits:

1. Decrease of saturated fat intake
    - Butter, Cheese, Meat, Whole Milk
2. Decrease of cholesterole intake
    - Eggs, Sea Food

Table 4. Criteria for Inference of Causality

	Cigarette smoking
1. Power of association	±
2. Dose/effect relationship	+
3. Temporal relationship (Factor → Event)	+
4. Same observations in several studies	±
5. Association is independent from other factors	+
6. Predictive power	±
7. Same observations in animal studies	±
8. There is a logical pathogenetic mechanism	?

3. Increase of polyunsaturated fat intake
    - Margarines, Oils (sunflower, soyabean), Fish
4. Increase of fiber intake
    - Vegetables, Fruits, Grains
    ----
5. Decrease of salt intake
6. Decrease of total calories

It is probable that part of the decrease of coronary mortality
in the USA, Australia, New Zealand, Israel, Finland and Belgium is
the result of slow-acting mass modification of food habits, acting
on serum lipids and blood pressure, and also a decrease in the
smoking epidemic. The individual approach is aimed at the modification of the risk profile in high risk individuals or in general in
the clinical setting. In the Belgian Heart Disease Prevention Project both approaches were used: face-to-face counselling in high
risk subjects (21% of the target population) and massmedia in the
rest of the population[7]. The coronary risk profile defined by means
of the multiple logistic risk function was significantly decreased
in the high-risk intervention group as compared to the control
group (Figure 1) throughout the trial although less so after six

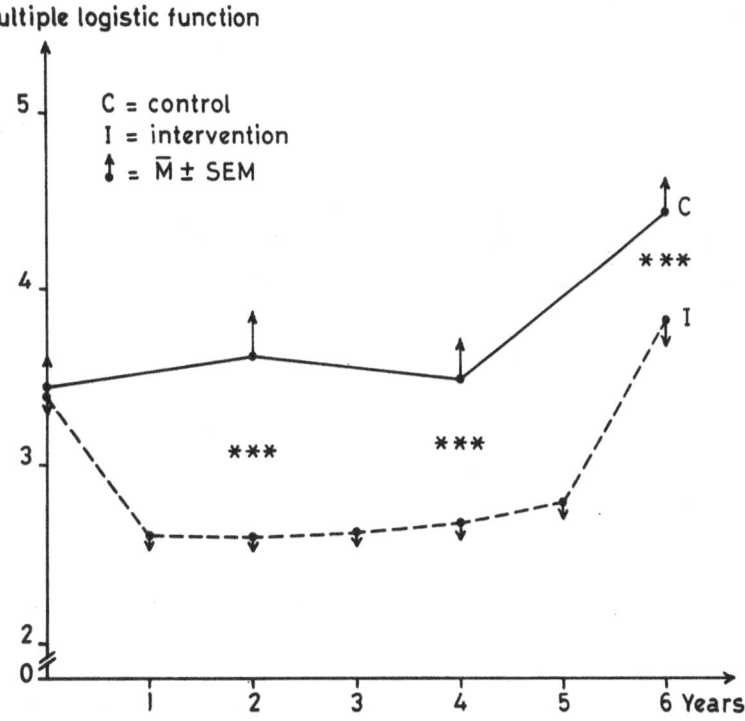

Fig. 1. Belgian heart disease prevention project. High risk groups.

years of follow-up. For the random samples of the whole groups,
where 79% had essentially received advice by means of booklets and
posters, the coronary risk profile was not permanently modified:
whereas at 2 and 4 years we observed a significant difference
between intervention and control groups, no such differences were
observed at 6 years. This was due to a lessening of influence on
blood pressure and smoking (Figure 2).

I would like to take one typical example, that of high blood
pressure. At initial screening 50% of all hypertensives (defined
as subjects having a mean systolic blood pressure ⩾160mmHg after 4
measures) were not in the high risk group. They were seen only
twice after the initial screening. The base-line study of hypertensive subjects showed 20.3% to be treated and well controlled,
15.7% treated and not well controlled, 50.8% never treated and
finally 13.2% temporarily treated (Table 5). After a follow-up of
6 years, 60% of initially non-treated hypertensives were still not
under treatment (never: 41.1%; temporarily: 18.7%)(Table 6). Who
is to be blamed? The patient or the physician? We were referring
throughout the whole 6 years all hypertensive subjects to their

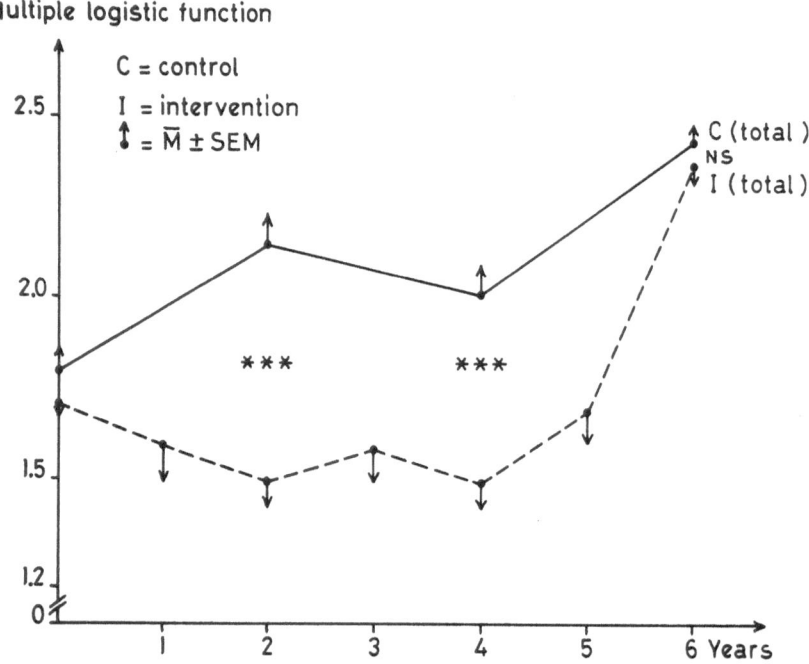

Fig. 2. Belgian heart disease prevention project. Total groups
         (or random samples).

Table 5. Belgian Heart Disease Prevention Project. Base-line Study of Hypertensive Subjects

	Executives (N=53)		White-collars (N=127)		Blue-collars (N=412)		Total (N=592)	
	N	%	N	%	N	%	N	%
Under Treatment (Not Controlled)	12	22.6	19	15.0	62	15.0	93	15.7
Under Treatment (Controlled)*	10	18.9	30	23.6	80	19.4	120	20.3
Non Treated:								
- Never	29	54.7	63	49.6	209	50.7	301	50.8
- Temporarily	2	3.8	15	11.8	61	14.8	78	13.2

* First systolic blood pressure <160mmHg.

Table 6. Belgian Heart Disease Prevention Project. Treatment Status at End-screening of Initially Non-treated Hypertensive Subjects

	Executives (N=19)		White-collars (N=45)		Blue-collars (N=150)		Total (N=214)	
	N	%	N	%	N	%	N	%
Under Treatment (End-screening)	7	36.8	21	46.7	58	38.6	86	40.2
Non Treated:								
– Never	7	36.8	17	37.7	64	42.7	88	41.1
– Temporarily	5	26.3	7	15.5	28	18.7	40	18.7

physicians as we were not allowed to prescribe drug treatment. At base-line we did not observe any significant difference in psychological traits or behaviors between treated and non- or temporarily treated subjects (Table 7). The same holds true when adjusting for social class except for blue-collars where age, level of systolic blood pressure, extroversion scale and job-involvement are significant discriminators (Table 8).

Treatment status at end-screening was only related with extroversion scale and level of systolic blood pressure at entry (Table 9). When adjusting for social class, age and level of systolic blood pressure are significant discriminators in white-collars and executives while extroversion is for blue-collars (Table 10). We got the firm impression that in the Belgian Heart Disease Prevention Project the major determinant of permanent treatment was the patients' physician although bad compliance (related with introversion?) can not be easily dismissed.

When and where should preventive advice be given? For mass intervention, prevention could start in childhood and adolescence at school and in adults by means of mass-media and in the army. Individual intervention could be done at the work place like in the Belgian Heart Disease Prevention Project by the factory physician and of course by the general practitioner. Concerning this last point I would remind you that during Prof. Denolin's chairmanship at the European Society of Cardiology a Working Group was convened in order to write a booklet for the practising physician[8].

Let us now summarize the published randomized trials concerning the primary prevention of coronary heart disease.

Table 7. Belgian Heart Disease Prevention Project. Hypertensive Subjects at Base-line Screening

	Treated (N=94)	Non-treated (N=384)	P.
Age	51.7	50.0	N.S.
Systolic Blood Pressure (first measure)	188.1	183.2	<0.01
S.H.-E.P.I. A Scale	12.1	12.0	N.S.
B Scale	5.1	5.0	N.S.
N Scale	8.8	7.4	N.S.
E Scale	10.9	11.5	N.S.
J.A.S.-AB	-4.9	-5.6	N.S.
J.A.S.-S	-8.0	-8.5	N.S.
J.A.S.-J	-13.3	-13.0	N.S.
J.A.S.-H	-0.4	-1.3	N.S.

Table 8. Belgian Heart Disease Prevention Project. Hypertensive Subjects
at Base-line Screening

| | Executives + White-collars (N=140) | | | Blue-collars (N=332) | | |
	Treated (N=31)	Non-treated (N=109)	P.	Treated (N=62)	Non-treated (N=270)	P.
Age	48.2	49.4	N.S.	53.5	50.2	<0.001
Systolic Blood Pressure (first measure)	187.0	183.7	N.S.	189.0	183.2	<0.01
S.H.-E.P.I. A Scale	12.2	12.3	N.S.	12.0	11.9	N.S.
B Scale	5.2	4.9	N.S.	5.0	5.1	N.S.
N Scale	9.0	7.6	N.S.	8.7	7.2	N.S.
E Scale	11.7	10.9	N.S.	10.5	11.7	<0.02
J.A.S.-AB	0.2	-2.7	N.S.	-8.0	-6.8	N.S.
J.A.S.-S	-3.3	-4.5	N.S.	-10.8	-10.2	N.S.
J.A.S.-J	-8.0	-9.5	N.S.	-16.4	-14.5	<0.02
J.A.S.-H	0.9	-0.9	N.S.	-1.2	-1.4	N.S.

Table 9.   Belgian Heart Disease Prevention Project. Psychological
           Base-line Factors in Relation with Treatment Status of
           Hypertensive Subjects at End-screening

	Treated (N=86)	Non-treated* (N=134)	P.
Age	49.3	48.3	N.S.
Systolic Blood Pressure	185.8	180.0	<0.01
(first measure)			
S.H.-E.P.I. A Scale	12.0	12.1	N.S.
B Scale	4.9	5.0	N.S.
N Scale	7.0	7.3	N.S.
E Scale	10.9	11.8	<0.02
J.A.S.-AB	-6.6	-5.3	N.S.
J.A.S.-S	-9.2	-8.7	N.S.
J.A.S.-J	-12.3	-13.5	N.S.
J.A.S.-H	-1.5	-1.0	N.S.

* Never or temporarily treated.

A.   UNIFACTORIAL PREVENTION TRIALS

1.   Trials acting on serum cholesterol

     Four trials, all dating back to the late fifties, had decreased
serum cholesterol (reduction range 9 - 12.7%) by dietary means:
decrease of saturated fat and cholesterol intake, increase of poly-
unsaturated fat[9-12]. In all four trials a significant difference
between experimental and control groups in terms of coronary or
cardiovascular events was observed with some pecularities for each
of them:

     1) In the Dayton et al.[10] trial fatal atherosclerotic events
        were significantly decreased only in the younger subjects,
        aged 54-65 at start. Besides total mortality was not de-
        creased in the experimental group.
     2) In the Turpeinen et al.[12] trial, coronary mortality was
        significantly reduced in males but not in females, while
        again neither in males nor in females was total mortality
        significantly reduced in the experimental group as compared
        to the control group.
     3) In the Christakis et al.[9] trial coronary morbidity was
        significantly decreased in the experimental group whereas
        coronary mortality was not.
     4) In the Frantz et al.[11] trial no significant differences were
        observed in the morbidity and mortality among older men or
        among women in the dietary treatment and in the control
        group. However, in the younger 1,192 men randomized to the
        treatment group and 1,106 men randomized to the control

Table 10. Belgian Heart Disease Prevention Project. Psychological Base-line Factors in Relation with Treatment Status of Hypertensive Subjects at End-screening

| | Executives + White-collars (N=65) | | | Blue-collars (N=152) | | |
	Treated (N=28)	Non-treated* (N=37)	P.	Treated (N=58)	Non-treated* (N=94)	P.
Age	50.1	47.2	<0.02	48.9	49.1	N.S.
Systolic Blood Pressure (first measure)	185.9	178.6	<0.01	185.8	180.8	N.S.
S.H.-E.P.I. A Scale	12.0	12.4	N.S.	12.0	12.0	N.S.
B Scale	5.2	4.6	N.S.	4.8	5.2	N.S.
N Scale	7.0	6.9	N.S.	7.0	7.5	N.S.
E Scale	10.8	11.3	N.S.	10.9	12.0	<0.03
J.A.S.-AB	-3.8	-4.9	N.S.	-7.9	-5.6	N.S.
J.A.S.-S	-5.0	-7.5	N.S.	-11.4	-9.4	N.S.
J.A.S.-J	-8.1	-10.4	N.S.	-14.4	-14.7	N.S.
J.A.S.-H	-1.9	-0.9	N.S.	-1.3	-1.0	N.S.

* Never or temporarily treated.

group, all below age 50, significant differences were found
for both coronary events and total mortality.

All four trials were heavily criticized in terms of design or
randomization or follow-up procedures. In fact, all four had a
weak statistical power as number of participants in each of them
was small (except maybe for the Frantz et al. trial). In fact they
paved the way for the multi-factorial trials which started in the
early seventies. A fifth trial acted on high serum cholesterol by
means of a drug: the Clofibrate trial was initiated by W.H.O. in
1965.[13] Population consisted of males aged 30-59 years living in
Edinburgh, Prague and Budapest. Distribution of serum cholesterol
in 15,000 subjects yielded 5,000 in the upper third from whom half
was randomly affected to the experimental group receiving 1.6g/day
clofibrate, whereas the control group received capsules containing
olive oil. In the lower third of the distribution, half of the
subjects (low-cholesterol group) received also the same placebo of
olive oil. The results of this trial were rather startling. Whereas
total C.H.D. and non fatal M.I. incidence were significantly reduced
in the experimental group, fatal M.I. was not and all causes of mortality were significantly higher in the experimental group (as was the
non cardiovascular mortality). Clofibrate reduced serum cholesterol by
9%. The excess of death was due to liver - gallbladder - small and
large intestine diseases. An excess of gallstone formation with
consequent cholescystectomies was also observed in the experimental
group. After a mean observation of 9.6 years, there were 25% more
deaths in the clofibrate treated group, than in the control group II
(high serum cholesterol)[14]. The treated group had more deaths from
C.H.D., cancer, stroke and major diseases[14]. Explanation of the
excess mortality is not apparent but, one of the co-directors of
this trial stated: "Clofibrate cannot be recommended as a lipid-
lowering drug for the community-wide prevention of coronary heart
disease".[15]

2. Trials acting on cigarette smoking

One published trial can be referred here, that of Rose et al.
[16] which was initiated in 1968 in 1,445 male smokers. At one year,
51% of the experimental group reported being ex-smoker. Mortality
follow-up was 10 years. A non statistically significant reduction
of 18% in coronary mortality was observed in the intervention group
whereas the reduction in lung cancer was 23%; deaths from non-lung
cancers were significantly higher in the intervention group so that
total mortality was about equal in both groups.

3. Trials acting on high blood pressure

Here six randomized trials can be referred to, dividing the
V.A. Cooperative Study in two substudies according to initial level

of diastolic blood pressure[17-22]. We will briefly summarize them:

The V.A. Cooperative Trial started in 1963. The first publication concerned 143 males[17]. Mean ages were respectively 51.4
(control) and 50 (experimental). Administered drugs were: hydrochlorothiazide, reserpine and hydralazine. In the 143 subjects with
entry diastolic blood pressure between 115 and 129mmHg, the trial
had to be stopped early because of a significant excess of cardiovascular events (mortality and morbidity) in the control group,
directly related to hypertension. Then in 380 males with entry diastolic blood pressure between 90 and 114mmHg, total mortality was
reduced by 50% in the experimental group, cardiovascular and C.H.D.
mortality were reduced whereas non significant differences were observed for total C.H.D.[18]. This trial was in fact not meant to show
differences in C.H.D. incidences as it was stopped when differences
in other target end-points were reached.

The U.S. Public Health Cooperation Study Group initiated a
randomized trial in 389 subjects, with mild hypertension, mean age
44, in 1964.[19] Administered drugs: chlorothiazide and reserpine.
No differences in cardiovascular and C.H.D. mortality were observed,
nor was there any difference between the two groups in total C.H.D.
incidence.

The Australian Therapeutic Trial in Mild Hypertension[20] was
initiated in 1973 in 3,427 subjects, mean age 50.4 years. Definition of hypertension: diastolic blood pressure between 95 and 105
with systolic blood pressure under 200mmHg. Administered drugs:
first chlorothiazide, then methyldopa or propranolol or pindolol was
added, then (eventually) clonidine or hydralazine. A reduction of
29% in total mortality was observed in the experimental group. Important reductions in cardiovascular and C.H.D. mortality were also
observed whereas total C.H.D. hard end-points were not significantly
different in both groups.

The Hypertension Detection and Follow-up Programme was initiated in the U.S. in 1979 [21]: 10,940 subjects (males + females) aged
30-59 years were randomized by strata of diastolic blood pressure
(90-104; 105-114; 115+) into a Stepped Care and Referred Care Group.
Administered drugs: diuretics, antidrenergics, vasodilatators.
After a 5 year follow-up, a significant reduction in total mortality
of 17% was observed. In the subgroup, with entry pressures between
90 and 104mmHg, a reduction of total mortality of 20% was achieved.
Preliminary results concerning specific mortalities are favorable:
total cardiovascular - 18.8%; total C.H.D. mortality - 15%; cerebrovascular mortality - 45%.

The Oslo Study initiated in 1973 a trial in 785 males aged
40-49 at entry[22]. Administered drugs: hydrochlorothiazide, methyldopa
or propranolol. A significant reduction in cardiovascular mortality

+ morbidity (all events) has been observed whereas the reduction in
total C.H.D. hard event was non significant.

In summary, most of these trials show a definite benefit in
terms of decrease of cerebrovascular mortality and/or morbidity in
the experimental group as compared to the control group. For coronary mortality and/or morbidity things are less clear although both
the Australian Therapeutic Trial and the Hypertension Detection and
Follow-up Programme has shown a definite benefit in terms of coronary
mortality.

B.   THE MULTIFACTORIAL TRIALS

1.   The Oslo Study

This was initiated in 1973 and is aimed at the reduction of
both serum cholesterol and smoking in high risk (H.R.) subjects;
1232 H.R. subjects were randomized and in the experimental group
subjects were given face-to-face counselling. Published results
show a reduction of 50% in total 5 year cardiovascular incidence in
the experimental group as compared to the control group[5].

2.   The North Karelia Project

This was established after a population petition to reduce
the extremely high C.V.D. rates in North Karelia[23]. It was initiated
in 1972.[6]
         Principles: 1. Controlled but not randomized.
                     2. Multifactorial aiming at life-style modification.

These factors were reduced in North Karelia as compared to the
reference area. Reduction in serum cholesterol was 4.1% in males
and 1.2% in females. For systolic blood pressure, figures were
respectively -3.6% and -4.8%. For prevalence of smokers -1.3% and
-0.7%. The coronary risk profile (computed by means of the multiple
logistic function) was reduced by 17.4% in males and 11.5% in females.
Important reductions in cardiovascular coronary and stroke mortalities
were observed. But as some reductions were also observed in the
reference area, the differences were not statistically significant.
This trial has shown that it is possible to modify the coronary risk
profile and consequently C.H.D. incidence in an area of over 200,000
citizens by means of an integrated approach using mass-media and
individual counselling.

In summary, while the issue of primary prevention of C.H.D.
should no more be questioned in 1982, several major questions
remain open to discussion, the most important of those being the
psycho-social factors involved in the initiation of ill-behaviors
and possibility of their long-term modification.

REFERENCES

1. G. Rose, P. J. S. Hamilton, H. Keen, D. D. Reid, P. MacCartney,
    R. J. Jarrett, Myocardial ischaemia, risk factors and death
    from coronary heart disease, Lancet, i:105 (1977).
2. A. M. Telford, C. Wilson, Trial of heparin versus atenolol in
    prevention of myocardial infarction in intermediate coronary
    syndrome, Lancet, i:1225 (1981).
3. A. Keys, Seven Countries, A multivariate Analysis of Death and
    Coronary Heart Disease, Harvard University Press, Cambridge,
    Massachusetts and London, England (1980).
4. The Multiple Risk Factor Intervention (MRFIT), A national study
    of primary prevention of coronary heart disease, JAMA., 235:
    825 (1976).
5. I. Hjermann, K. V. Byre, I. Holme, P. Leren, Effect of diet and
    smoking intervention on the incidence of coronary heart
    disease, Lancet, ii:1303 (1981).
6. J. T. Salonen, P. Puska, H. Mustaniemi, Changes in morbidity and
    mortality during comprehensive community programme to control
    cardiovascular diseases during 1972-7 in North Karelia, Brit.
    Med. J., 2:1178 (1979).
7. M. Kornitzer, G. de Backer, M. Dramaix, C. Thilly, The Belgian
    Heart Disease Prevention Project: Modification of the Coronary
    risk profile in an industrial population, Circulation, 61:18
    (1980).
8. European Society of Cardiology, Preventing Coronary Heart Disease;
    A guide for the practising physician, Van Gorcum Assen, The
    Netherlands (1978).
9. G. Christakis, S. H. Rinzler, M. Archer, A. Kraus, Effect of the
    anti-coronary club program on coronary heart disease: risk
    factor status, JAMA., 198:129 (1966).
10. S. Dayton, Rationale for use of lipid-lowering drugs, Federation
    Proceedings, 30:849 (1971).
11. I. D. Frantz, E. A. Dawson, K. Kuba, E. R. Brewer, L. C. Gatewood,
    G. E. Bartsch, The Minnesota Coronary Survey: effect of diet
    on cardiovascular events and deaths (Abstract), Circulation,
    51 and 52, suppl. II, p. 4 (1975).
12. O. Turpeinen, M. Miettinen, M. J. Karvonen, P. Roine, M. Pekka-
    rinen, E. J. Lehtosuo, P. Alivirta, Dietary prevention of
    coronary heart disease: long-term experiment. I. Observations
    on male subjects, Am. J. Clin. Nutr., 21:255 (1968).
13. M. F. Oliver, J. A. Heady, J. N. Morris, J. Cooper, A cooperative
    trial in the primary prevention of ischaemic heart disease
    using clofibrate, Brit. Heart J., 40:1069 (1978).
14. M. F. Oliver, J. A. Heady, J. N. Morris, J. Cooper, W.H.O. Co-
    operative trial on primary prevention of ischaemic heart
    disease using clofibrate to lower serum cholesterol: Mortality
    follow-up, Lancet, ii:379 (1980).
15. M. F. Oliver, Cholesterol, coronaries, clofibrate and death,
    N. Engl. J. Med., 299:1360 (1978).

16. G. Rose, P. J. S. Hamilton, L. Colwell, M. J. Shipley, A random-
    ized controlled trial of anti-smoking advice: 10 year results,
    J. Epidem. Com. Hlth., 36:102 (1982).
17. Veterans Administration Cooperative Study Group on Antihyper-
    tensive Agents, Effects of treatment on morbidity in hyper-
    tension: I. Results in patients with diastolic blood pressure
    averaging 115 through 129mmHg, JAMA., 201:1028 (1967).
18. Veterans Administration Cooperative Study Group on Antihyper-
    tensive Agents, Effects of treatment on morbidity in hyper-
    tension: II. Results in patients with diastolic blood pressure
    averaging 90 through 114mmHg, JAMA., 213:1143 (1970).
19. W. McFate Smith, Treatment of mild hypertension: Results of a
    ten-year intervention trial, Circ. Res., 40 Suppl. I., 98
    (1977).
20. The Australian Therapeutic Trial in Mild Hypertension, Report
    by the Management Committee, Lancet, i:1261 (1980).
21. Hypertension Detection and Follow-up Program Cooperative Group,
    Five-year findings of the hypertension detection and follow-
    up program: I. Reduction in mortality of persons with high
    blood pressure, including mild hypertension, JAMA., 242:2562
    (1979).
22. A. Helgeland, Treatment of mild hypertension: a five year con-
    trolled drug trial, The Oslo Study, Am. J. Med., 5:725 (1980).
23. J. Tuomilehto, The most recent lesson from community control of
    cardiovascular diseases, Acta Cardiol., 35:251 (1980).

PRIMARY PREVENTION TRIALS OF CORONARY HEART DISEASE

A. Menotti

Laboratory of Epidemiology
Istituto Superiore di Sanità
Rome, Italy

The observational epidemiological studies on coronary heart disease (CHD) conducted along the last 30 years all over the world have identified the so-called coronary risk factors whose discriminant power in the prediction of the disease is well known[1-22]. Although coronary risk factors cannot be automatically considered as causes of the disease, a number of circumstantial evidences from epidemiology and other disciplines have suggested their important role in the development of morbid events and therefore it became obvious the attempt to set up preventive trials based on the so-called hypothesis of the reversibility of coronary risk. The hypothesis assumes that, starting from given levels of factors (corresponding, when taken in pool, to the overall estimated risk) an artificially induced reduction of such levels should correspond, within a reasonable period of time, to a "proportionate" reduction of CHD incidence in the population (Fig. 1). The expected decrease of incidence should be similar to the observed difference between two populations having the risk factors at the same levels recorded at entry and at the end of a trial in an experimental population. A trial of this type would, of course, be run with a reference untreated population, as control. Most of the theory and practice in this field has been developed in relationship to the primary prevention of CHD although many studies of secondary prevention have been carried out in post-infarction patients with the purpose to reduce late complications, relapses and mortality.

The chances of success of a primary preventive trial of CHD are bound to a number of conditions, the following being the main ones:
(a) the factors are modifiable;
(b) the factors are at least indirectly causal and not only indicators of future risk;

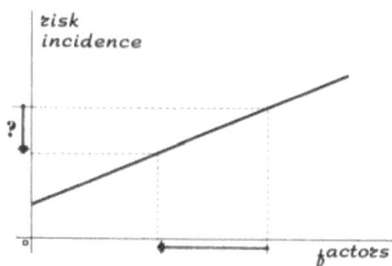

Fig. 1. The theoretical structure of the hypothesis of reversibility
of coronary risk.

(c) the induced changes of factors are large enough and persistent;
(d) the eventual damage already accumulated is not too advanced and
    possibly suitable to regression;
(e) the latency between changes of risk factors levels and change
    of the actual risk is not too long;
(f) the risk factors changes do not induce other diseases beside
    preventing CHD;
(g) the amount of eventual incidence reduction is large enough to be
    measurable.

An alternative to the hypothesis of the reversibility of coronary risk is that of the prevention of risk, or pre-primary prevention, or primordial prevention of CHD[23]. It assumes that, owing to
the long term induction of atherosclerosis, and to the low levels
(or even absence, like for smoking habits) of all main risk factors
in children, the ideal preventive action would be to avoid the
increase, or to put a brake to the increasing levels of risk factors
with ageing. Although this might be considered the best approach to
primary prevention, it is unsuitable for an experimental controlled
trial.

Referring to a classical primary preventive trial of CHD, its
basic structure is not different from that of any other clinical
trial[24]. Two groups of people randomly allocated to either control
or treatment are followed-up for a given period of time in view of
detecting a possible different incidence rate between the two groups

as a function of the effectiveness of the preventive treatment in
the intervention group as compared to no treatment in the controls.
Such an approach presents a number of problems since incidence rate
of CHD is rather low (if compared for example with the recovery rate
of a disease in a typical therapeutical trial) and the sample size
must be consequently quite large. Moreover the avoidance of contamination between treatment and control groups is not necessarily
easy, and last, but not least, the so-called control groups are not
usually comparable to guinea-pigs in a cage, since they are free to
ask for medical and hygienic advice from the existing health services.
This means that the control groups are usually to be considered
reference groups or even referred-care groups. The trial, in this
case, is directed to obtain a net gain of risk factor changes in
the treated over the control groups. It must also be recognized
that, if a population is involved in a trial, what has to be changed
is not only the extremely high levels of risk factors, but mainly
the whole distribution curve which should be moved to the left,
even if only a relatively small degree. This is the essential condition for hoping in a measurable decrease of incidence if the
hypothesis holds.

A critical problem, already mentioned above, to be solved for
the organization of a preventive trial is the sample size estimate.
A good model has been developed by the Study Group of the American
Diet Heart Study[25], which takes into account the following variables:
(a) instantaneous incidence rate for control group;
(b) instantaneous drop-out rate for experimental group;
(c) latency before achieving maximum effect;
(d) proportionate expected reduction of incidence in experimental
    group, as a function of estimated risk factor changes;
(e) alpha error (probability of rejecting the null hypothesis when
    it is true);
(f) beta error (probability of accepting the null hypothesis when
    it is false).
Suitable tables have been constructed for easy estimate of the
sample size, which is usually very large.

Another important question to be answered before starting a
trial concerns the amount of expected incidence reduction of CHD
as a function of risk factors changes, which can be reasonably
achieved. The assumption is, of course, only theoretical and represents a byproduct of the multivariate estimate of CHD as a function
of risk levels, produced by observational studies. The expected
reduction starts from the hypothesis of zero time latency and full
regression of possible damages. An example of estimates of this
type is reported in Table 1 where Italian epidemiological data
treated with the multiple logistic function have been used[26]. It
becomes clear that relatively small changes of mean levels of risk
factors may produce measurable decreases of incidence, whereas
large changes could provide exceptionally large incidence reduction.

Hypotheses have to be made also about the possible types of results of a preventive trial when trying to correlate the observed changes of risk factors with the observed differences in incidence, by comparing treatment with control groups. Table 2 tries to make a number of hypotheses whose interpretation is sometimes very hard. Failure to change the risk factors suggests a unfeasibility of prevention or a non-effectiveness of the tools employed for intervention. Concordant changes of risk factors and incidence are anyhow not against the theory which links the two phoenomena; whereas discordant changes put a lot of question marks. Concordance in decrease, and even better when the amount of decrease of incidence is in line with that expected from risk factors changes, is the optimal and desired result.

There are different types of trial structures, which involve different kinds of organization and different types of questions to be answered. Three prototypes can be mentioned:
(1) Individual intensive intervention is limited to very high risk subjects selected from a population screening against no treatment or referred care of a control group of similar nature. The allocation of individuals to treatment of control is made randomly. The trial is able to answer the question of the prevention of CHD only in very high risk subjects, ignoring the remaining fraction of the

Table 1.   Theoretical Expected Reduction of CHD Incidence for Defined Changes of some Risk Factors in the General Population.   Computed from Italian Data of the Seven Countries Study[26].

Changes of Factors %	Incidence Reduction %
CHOL 5	10
CHOL 10	19
CHOL 15	25
CHOL 20	34
CHOL 25	41
SBP 5	13
SBP 10	25
SBP 15	35
CIG 50	10
CIG 100	22
CHOL 5; SBP 5	22
CHOL 10; SBP 5	31
CHOL 10; SBP 10	40
CHOL 5; SBP 5; CIG 50	30
CHOL 20; SBP 10; CIG 100	62
CHOL 25; SBP 15; CIG 100	72

population, although the high risk group might produce a large proportion of the overall incidence cases in the population. Typical
trials of this type are the Multiple Risk Factors Preventive Trial
in the USA[27], the Oslo Study[28,29], the Gotenburg Study[30,31].
(2) Whole groups (either demographic or occupational) are randomly
allocated in pairs to either control or treatment. In the latter,
intervention is provided individually and intensively to subjects
located in the upper levels of a risk score (15 to 30%), whereas
general mass health education is provided to the remaining section
of the treatment groups. In this way the total treatment population
is exposed to some kind of intervention, while the controls are left
to their usual health care. The answer on the possibility to prevent
CHD is therefore extended to the whole population, but the allocation
of groups instead of individuals does lower the power of the trial.
The prototype of this kind of trial is the WHO European Multifactor
Preventive Trial of CHD, where pairs of "factories" have been allocated to either control or treatment[32,33].
(3) Another type of trial, definitely looser and sometimes simply
called demonstration project, involves very large general population
groups to whom only general mass education and re-organization of
health services in view of prevention are offered, and incidence
data are collected not individually but through a registration system. Another population group acts as reference. A typical study
of this type is the North Karelia Project[34].

Table 2. Possible Findings in a Primary Preventive Trial

Risk Factor Change in Treatment vs Control Groups	Incidence in Treatment vs Control Groups	Conclusions
=	=	Unfeasible ?
=	Up	Unfeasible ??
=	Down	Chance ?
Up	=	Unfeasible ?
Up	Up	Wrong time ?
Up	Down	! !
Down	=	Long latency
Down	Up	! ! !
Down	Down	O. K.

    In the recent story of primary prevention of CHD, it is common
to identify two generations of trials:
(1) the first generation trials have been conducted and concluded
mainly in the sixties; usually they were unifactorial and frequently
they have been criticized for a number of problems including sample
selection and size, types of controls, etc. They have provided some
suggestive but not necessarily conclusive results.

Table 3. Some Examples of Primary Preventive Trials of CHD, First Generation

Study	Action on Factor(s)	Results
Anticoronary Club[35]	Diet on cholesterol	Positive, suggestive
Finnish Mental Hospitals[36]	Diet on cholesterol	Positive, suggestive
Veterans Administration Dayton[37]	Diet on cholesterol	Positive, suggestive
National Diet Heart[25]	Diet on cholesterol	Positive, suggestive
Veterans Administration Freis[38,39]	Drugs on hypertension	Positive for strokes
		Negative for infarction
Whitehall[40]	Advice on smoking	Unknown

(2) The second generation trials have been conducted in the seventies and only some of them have been concluded. Most were of multifactorial type of intervention, with higher probabilities to get a final answer; they had larger samples, proper control groups and so far have produced some important results in line with the theory. In all cases the intervention has been pointed, either unifactorial or multifactorial, to a limited number of major risk factors, i.e. serum cholesterol to be lowered by diet (or drugs); high blood pressure, to be lowered by the use of anti-hypertensive drugs of various type; smoking habits to be attacked by counselling of different kinds. Only marginally have other factors been considered and treated. The intervention has been rather variable but some patterns have been common to all of the studies. A summary of the main studies of the two generations is reported in Tables 3 and 4. The merging picture is reasonably optimistic since most of the results seem to be in line with the hypothesis of the reversibility of risk and with the feasibility of an efficient intervention. Two large studies will disclose their - still secret - results in June and September 1982 (the WHO European Multifactor Preventive Trial of CHD and the American Multiple Risk Factor Intervention Trial, respectively) and two more are expected for 1983 (Whitehall Trial in Great Britain and the Gotenburg Trial in Sweden).

   In spite of that, problems are still open, which need much attention in the near future. Among them the following are identified as some major ones:

(a) the demonstration of a harmless effect of long term use of antihypertensive drugs;
(b) the demonstration of a harmless effect of substantial changes in the diet, including the increase of poly-unsaturated fat and the decrease of salt consumption;
(c) the demonstration that such results, obtained in research trials, can be transferred to large free living population groups;
(d) the identification of means for the non-pharmacologic treatment or prevention of high blood pressure;
(e) the feasibility of an intervention on children in order to brake the increase of the rising risk factors levels with ageing

All these problems are being tackled now by the so-called third generation of the preventive studies on CHD. Still, beyond these problems, further knowledge has to be acquired if the final aim points toward a possible, but so far unlikely, eradication of the disease.

Table 4. Some Examples of Primary Preventive Trials of CHD, Second Generation

Study	Action on Factor(s)	Results
Clofibrate Trial[41]	Drug on cholesterol	Partially positive on non-fatal CHD
Multiple Risk Factor Intervention Trial[27]	Diet on cholesterol Drugs on hypertension Advice on smoking	Some changes of factors No incidence data available
North Karelia Project[34]	General mass education Organization of services	Partially positive
WHO European Multifactor Preventive Trial of CHD[32,42]	Diet on cholesterol Drugs on hypertension Advice on smoking Diet on overweight Advice on sedentarity General health education	Some changes of factors No incidence data available
Lipid Clinics Research Programme[43]	Drugs on blood lipids	Unknown
Gotenburg Trial[30,31]	Diet on cholesterol Drugs on hypertension Advice on smoking	Reduced incidence of infarction in treated hypertension
WHO Community Control of Hypertension[44]	Drugs on hypertension	Unknown
Oslo Study[28,29]	Diet on cholesterol Advice on smoking Drugs on hypertension	Substantial reduction of infarction incidence, no effect on CVD
Hypertension Detection and Follow-up Programme[45]	Drugs on hypertension	Reduction of all causes mortality
Australian Hypertension Trial[46]	Drugs on hypertension	Reduction of all causes mortality Some reduction of CHD incidence

REFERENCES

1. D. Shurtleff, "Some characteristics related to the incidence of
    cardiovascular disease and death: Framingham Study, 18 year
    follow-up: Section 13", Dept. of Health, Education and
    Welfare, Publ. (NIH), n.74-599, Bethesda Md., (1974).
2. Pooling Project Research Group, Relationship of blood pressure,
    serum cholesterol, smoking habits, relative weight and ECG
    abnormalities to incidence of major coronary events: final
    report of the Pooling Project, J. Chron. Dis., 31:201 (1978).
3. A. Keys (ed.), Coronary heart disease in Seven Counties.
    Circulation, 41:Suppl. 1 (1970).
4. A. Keys, C. Aravanis, H. Blackburn, R. Buzina, B. S. Djordjevic,
    A. S. Dontas, F. Fidanza, M. J. Karvonen, N. Kimura, A.
    Menotti, F. Mohacek, S. Nedeljkovic, V. Puddu, S. Punsar,
    H. L. Taylor and F. S. P. Van Buchem, "Seven Countries. A
    multivariate analysis of coronary heart disease and death",
    Harvard University Press, Cambridge Mass. and London,
    England (1980).
5. L. Wilhelmsen, H. Wedel and G. Tibblin, Multivariate analysis
    of risk factors for coronary heart disease, Circulation,
    48:950 (1973).
6. P. Ducimetiere, J. L. Richard, F. Cambien, R. Rakotovao and
    J. R. Claude, Coronary heart disease in middle-aged Frenchmen:
    Comparison between Paris Prospective Study, Seven Countries
    Study and Pooling Project, Lancet, 1:1346 (1980).
7. U. Goldbourt, J. H. Medalie and H. N. Neufeld, Clinical myo-
    cardial infarction on a five-year period. III: A multivariate
    analysis of incidence: The Israel ischemic heart disease
    study, J. Chron. Dis., 28:217 (1975).
8. A. Keys, H. L. Taylor, H. Blackburn, J. Brozek, J. T. Anderson
    and E. Simonson, Mortality and coronary heart disease among
    men studied for 23 years, Arch. Int. Med., 128:201 (1971).
9. J. Truett, J. Cornfield and W. Kannel, A multivariate analysis
    of coronary heart disease in Framingham, J. Chron. Dis.,
    20:511 (1967).
10. S. Walker and D. B. Duncan, Estimation of the probability of
    an event as a function of several independent variables,
    Biometrika, 54:167 (1967).
11. A. Keys, C. Aravanis, H. Blackburn, F. S. P. Van Buchem, R.
    Buzina, B. S. Djordjevic, F. Fidanza, M. J. Karvonen, A.
    Menotti, V. Puddu and H. Taylor, Probability of middle-aged
    men developing coronary heart disease in five years, Circu-
    lation, 45:815 (1972).
12. A. Menotti, R. Capocaccia, S. Conti, G. Farchi, S. Mariotti,
    A. Verdecchia, A. Keys and S. Punsar, Identifying subsets of
    major risk factors in multivariate estimation of coronary
    risk, J. Chron. Dis., 30:557 (1977).
13. A. Menotti, S. Conti, S. Giampaoli, S. Mariotti and P. Signor-
    etti, Coronary risk factors predicting coronary and other
    causes of death in fifteen years, Acta Cardiol., 35:107 (1980).

14. A. Menotti and S. Conti, Relazioni tra il livello di rischio
    coronarico quale espresso da una selezione di 36 fattori ed
    il rischio di altre condizioni morbose, Ann. Ist. Sup.
    Sanità, in press (1982).
15. Italian Research Group of the Seven Countries Study, Incidence
    and prediction of coronary heart disease in two Italian rural
    population samples followed-up for 20 years, Acta Cardiol.,
    in press (1982).
16. T. Gordon, W. B. Kannel and M. Halperin, Predictability of
    coronary heart disease, J. Chron. Dis., 32:427 (1979).
17. T. Strasser, Atherosclerosis and coronary heart disease: the
    contribution of epidemiology, WHO Chron., 26:7 (1972).
18. P. N. Hopkins and R. R. Williams, A survey of 246 suggested
    coronary risk factors, Atherosclerosis, 40:1 (1981).
19. A. Keys, C. Aravanis, F. S. P. Van Buchem, H. Blackburn, R.
    Buzina, B. S. Djordjevic, A. S. Dontas, F. Fidanza, M. J.
    Karvonen, N. Kimura, A. Menotti, S. Nedeljkovic, V. Puddu,
    S. Punsar and H. L. Taylor, The diet and all causes of death
    rates in the Seven Countries Study, Lancet, II:58 (1981).
20. V. Puddu and A. Menotti, An Italian study of ischemic heart
    disease, Acta Cardiol., 24:558 (1969).
21. A. Menotti and V. Puddu, Epidemiology of coronary heart disease:
    A 10 year study in two Italian rural population groups, Acta
    Cardiol., 28:66 (1973).
22. G. Farchi, R. Capocaccia, A. Verdecchia, A. Menotti and A. Keys,
    Risk factors changes and coronary heart disease in an obser-
    vational study, Int. J. Epidem., 10:31 (1981).
23. G. Lamm, "The Cardiovascular Disease Programme of WHO in Europe",
    Public Health in Europe, 15, (annex 7), WHO, Regional Office
    for Europe, Copenhagen (1981).
24. R. Saracci, Controlled studies, in: "Health Care and Epidemi-
    ology", W. W. Holland and L. Karhausen, eds., H. Kimpton
    Publishers, London (1978).
25. National Diet-Heart Study, Research Group, the National Diet-
    Heart Study final report, Circulation, 37:suppl. 1 (1968).
26. A. Menotti, "La prevenzione della Cardiopatia Coronarica",
    Pensiero scientifico, Roma (1976).
27. Forum: The Multiple Risk Factor Intervention Trial (MRFIT):
    The methods and impact of intervention over four years,
    Preventive Medicine, 10, n.4:387 (1981).
28. A. Helgeland, Treatment of mild hypertension: a five year con-
    trolled drug trial, The Oslo Study, Am. J. Med., 69:725
    (1980).
29. I. Hjermann, K. Velve Byre, I. Holme and P. Leren, Effect of
    diet and smoking intervention on the incidence of coronary
    heart disease: Report from the Oslo Study Group of a random-
    ized trial in healthy men, Lancet, II, n.8259:1303 (1981).
30. L. Wilhelmsen, G. Tibblin and L. Werkö, A primary preventive
    trial in Gotenburg, Sweden, Prev. Med., 1:153 (1972).
31. L. Wilhelmsen, G. Berglund, R. Sannerstedt, L. Hansson, O.
    Andersson, R. Sievertsson and J. Wilkstrand, Effect of
    treatment of hypertension in the primary preventive trial,

Gotenburg, Sweden, Br. J. Clin. Pharmac., 7, suppl. 2:261S (1979).

32. World Health Organization European Collaborative Group: An international controlled trial in the multifactorial prevention of coronary heart disease, Int. J. Epidem., 3:219 (1974).

33. World Health Organization Collaborative Group: Multifactorial trial in the prevention of Coronary heart disease: 1) Recriutment and initial findings European Heart Journal, 1:73 (1980).

34. National Public Health Laboratory of Finland, "Community control of cardiovascular disease: the North Karelia Project", WHO Regional Office for Europe, Copenhagen (1981).

35. G. Christakis, G. Winslow, S. Jampel, J. Stephenson, G. Friedman, H. Fein, A. Kraus and G. Hames, The anti-coronary club: a dietary approach to the prevention of coronary heart disease; A seven year report, Am. J. Publ. Health, 56:299 (1966).

36. M. Miettinen, O. Turpeinen, M. J. Karvonen, R. Elosuo and E. Paavilainen, Effect of cholesterol lowering diet on mortality from coronary heart disease and other causes; A twelve-year clinical trial in men and women, Lancet, 2:835 (1972).

37. S. Dayton, M. L. Pearce, S. Hashimoto, W. J. Dixon and U. Tomiyasu, A controlled clinical trial of a diet high in unsaturated fat, Circulation, 40, suppl.2:1 (1969).

38. Veterans Administration Cooperative Study Group on Antihypertensive Agents, Effect of treatment on morbidity in hypertension: Results in patients with diastolic blood pressure averaging 115 through 129mmHg, JAMA, 202:1028 (1967).

39. Veterans Administration Cooperative Study Group on Antihypertensive Agents, Effects of treatment on morbidity in hypertension: Results in patients with diastolic blood pressure averaging 90 through 114mmHg, JAMA., 213:1143 (1970).

40. G. Rose and P. J. S. Hamilton, A randomized controlled trial of the effect on middle-aged men of advice to stop smoking, J. Epidemiol. Commun. Health, 32:275 (1978).

41. A cooperative trial in the primary prevention of ischaemic heart disease using clofibrate: Report from the Committee of Principal Investigators, Br. Heart J., 40:1069 (1978).

42. WHO European Collaborative Group, Multifactorial trial in the prevention of coronary heart disease: 2, Risk factor changes at two and four years, Europ. Heart J., in press (1982).

43. The Lipid Research Clinics Program, The coronary primary prevention trial: design and implementation, J. Chron. Dis., 32:609 (1979).

44. WHO, "Report of the WHO meeting on the Community Control of Hypertension Geneva 21-23 Nov. 1977", WHO Geneva (1978).

45. Hypertension Detection and Follow-up Program Cooperative Group, Five year findings of the hypertension detection and follow-up program: 1, Reduction in mortality of persons with high blood pressure, including mild hypertension, JAMA., 242:2562 (1979).

46. Report of the Management Committee: The Australian therapeutic trial in mild hypertension, Lancet, II, n.8259:1303 (1981).

CONTROLLING CORONARY RISK FACTORS IN THE COMMUNITY

Frederick H. Epstein

Institute of Social and Preventive Medicine
University of Zürich
CH-8006 Zürich

During the last 30 years, preventive cardiology has made great strides. Prior to the 1950's, the field hardly existed. Starting shortly after the end of the Second World War, a number of major, prospective epidemiological studies came into being and others started throughout the next 20 years. As a result, the risk factor concept developed. It is the key to preventive cardiology, implying that there are factors which precede that onset of clinical disease and, if controlled, will prevent or delay its onset. "Control," in this context, means not only the treatment of risk factors which are already present but to prevent, if ever possible, their development.

This discussion deals with the control of coronary heart disease (CHD) risk factors in the community. Such a topic would have hardly been chosen in the 1950's when, as mentioned before, CHD epidemiology became a major scientific activity. At that time, the main focus of concern was the detection of persons at high risk, i.e., screening for susceptibles. It was certainly recognized that such high-risk people must be found in the population at large, i.e., in the community, but it was not fully realized how difficult it is to mobilize healthy human beings to be "screened" and how even more difficult it is to have them take action to reduce their risk. "Community Medicine" as presently understood, is essentially a new term, overlapping of necessity with medical practice in the traditional sense but embracing first and foremost all the forces and influences which are concerned with the maintenance of health: the people themselves, the medical and allied health professions, the health agencies and authorities, the community and social organizations and the various administrative and political bodies responsible for preventive and curative care of the population. The resulting interactions give the community a dynamism of its own and make it more than the sum

41

of the people who live in it. Correspondingly, controlling risk
factors in the community involves more than the mere mechanics of
screening and organizing health education activities but the creation
of a spirit which will reinforce the actions taken and the messages
conveyed.

There are the principles and the practice of risk factor control
in the community. Here, the emphasis will be on the principles
which underlie the community approach. There are two fundamental
questions: (1) does CHD risk factor control, in fact, prevent or
delay clinical CHD? (2) how do community approaches to risk factor
control compare with other methods which aim at reducing the burden
of this disease? The first question, is, of course, basic and will
be discussed first: there is no point in reducing risk factors or
preventing their emergence if this will not lead to a reduction in
the incidence of the disease associated with them (i.e., is there a
cause-and-effect relationship?) To answer the second question,
"community approach" must be defined within the total framework of
preventive strategies, to be followed by estimates of the effectiveness of various strategies.

Having decided on the proper strategies toward risk factor
control, it must be asked how they can be put into effect and to
good use. Unless people can be convinced that their health is ultimately their own responsibility, no amount of health education can
possibly work. However, in addition to the responsibility of the
individual, the society has grave responsibilities as well. In fact,
the individual has little chance to act responsibly in an environment
which is indifferent or hostile to matters of public health. There
is a duty, therefore, both on the part of the individual and the
society. Both are equally important! This review will close with
a brief discussion of these issues.

Is Risk Factor Control Effective?

Risk factor control in the community applies to two kinds of
people: (1) those who are still clinically healthy and who are,
therefore, in need of "primary prevention" and (2) patients with
clinically manifest coronary heart disease who require "secondary
prevention". In each case, two questions arise: (1) can risk
factors be adequately lowered? (2) does such lowering result in a
reduction of disease incidence? Since risk factor lowering depends,
apart from the community facilities and the prevailing attitudes
toward prevention, on the motivation of the physician to advise
preventive treatment and the motivation of the person concerned to
accept his advice, the situation differs fundamentally according to
whether disease is already present or not. In someone who is still
overtly healthy, motivation toward preventive action will, in general,
be less strong than in a patient with disease whose doctor will take

it for granted that he should be treated and who himself wants treatment. This is not to say that the treatment of patients is always
satisfactory, as the problem of "compliance" illustrates, and it
must also be clear that drug treatment is more likely to be followed
than recommendations to change life styles like diet, smoking habits
or patterns of regular exercise. No attempt is being made here to
present a systematic analysis of all these difficult and important
approaches to preventive treatment before and after the onset of
illness. Instead, the issues of concern and practical importance
have been raised, in order to give a picture of the different considerations entering into disease prevention on the community level
and some of the questions which should be faced.

     With regard to primary prevention, it must be asked first if it
is possible to achieve adequate risk factor lowering in large groups
of people. Even though it is known that specific dietary changes
will lower serum cholesterol levels, it is not a priori obvious to
what degree such lowering can be effected in the population groups.
Similarly, it must be tested how much blood pressure lowering is
possible under everyday circumstances and to what extent people will
follow anti-smoking advice. Concerning these three major risk factors
serum cholesterol, blood pressure and smoking, a review of the studies
which have been carried out to date indicates that risk factor lowering in the general population and in special industrial or high-risk
groups is, indeed, possible[1]. The degree of effectiveness varies,
of course, with the type of population and the intensity of the
health education effort. In general, as would be expected, the
average percentage lowering for the three main risk factors is greater
for high-risk persons than the population at large. Most important,
where an assessment in terms of a multiple logistic "risk-score" has
been made, taking simultaneous account of all three risk factors,
this score is lowered by around 30% which would signify theoretical
reduction in disease risk by the same percentage.

     How much reduction in disease incidence has actually been
achieved in the primary prevention trials carried out to-date? Five
trials in which the preventive measure was reduction of serum cholesterol showed a lowering of coronary heart disease incidence around
25-50%[1]. Six primary prevention studies through treatment of hypertension similarly indicated a benefit (difference in coronary heart
disease incidence between treated and reference groups) ranging between 25 and 50%[1]. All of these intervention trials have methodological short-comings which have been repeatedly pointed out. Nevertheless, it is very unlikely that their uniformly positive results
are all misleading and fictitious! Before the end of 1982, the
results of the two largest and best-designed primary intervention
trials (the United States MRFIT trial and the WHO-coordinated multifactorial trial in 5 European countries) are expected and awaited
with a sense of much anticipation. Meanwhile, it must always be
kept in mind that the evidence for the effectiveness of primary

prevention does not rest on the findings of preventive trials alone but is supported by the results of clinical research, experimental studies, pathological observations and many descriptive epidemiological investigations[1].

Before leaving the question if prevention "works", brief mention must be made of secondary prevention. Community medicine cannot confine itself only to primary prevention, notwithstanding its paramount importance. The community includes people with disease in whom progression of illness must be prevented. Concerning preventive cardiology, a good deal can be done for survivors from myocardial infarction and patients with angina pectoris. Two recent reviews are available[2,3]. Apart from medical treatment, by-pass surgery and transluminal angioplasty have most likely become established procedures for selected patients. In order to lessen the burden of cardiovascular disease in the community, people in whom primary prevention (i.e., prevention of premature disease) has failed must be protected from further damage. On the basis of present knowledge, there has been progress in secondary prevention which has at least contributed to the decline of coronary heart disease mortality in the last decade in a country like the United States[4].

Preventive Strategies in the Community

Having accepted that prevention is effective in lowering coronary heart disease risk through reduction in risk factors, it must be asked how preventive action can be carried out most efficiently on the community level. Some 10 years ago, it was pointed out that a strategy of prevention aimed only at high-risk groups (called, at the time, the "clinical-individual approach") will have an important but not major impact on the total coronary heart disease burden in the community and that the "community approach" is required in addition, shifting the distribution curve of risk factors in the population toward the left[5]. The community approach implies that preventive measures will involve not only persons in the upper range of the risk factor distribution but the whole population, whether or not risk factors are in the elevated range. Quite recently, an eloquent and impressive appeal has been made for the community approach, now termed "mass strategy", as contrasted with the "high-risk strategy", corresponding to the clinical-individual approach[6]. The two new terms are now becoming established and will serve well. The older terms were chosen deliberately to introduce the concept of the "community" on the one hand and the key role of the practicing physician in what is now called the "high-risk strategy" on the other because it involves the traditional contact between the individual patient and his doctor on the clinical rather than the community level. The practicing physician is certainly playing a part in the mass strategy as well but, as pointed out in the introduction, the entire community organization must be called upon to mobilize the

mass of the people toward creating a structure and a climate in
which the mass strategy can succeed.

It is possible to make some calculations to estimate the likely
effectiveness of the two approaches. Taking the risk factor blood
pressure first (Table 1), using Framingham data as a base[7], let us
see the effect on the incidence of myocardial infarction and cardiac
death of shifting the entire distribution curve of systolic blood
pressure to the left, from a mean of 138mmHg to 128mmHg. The incidence corresponding to each 10mm blood pressure range is given in
the first two columns, and the respective number of men in the third
column. From this information, the number of "Events" can be calculated (4th column). Let us now assume that the whole distribution
is shifted 10mmHg to the left so that there are now 38 + 132 = 170
men with pressures under 110mmHg, instead of the original 38, re-
peating this process all down the line to get to a new number (N')
of men in each blood pressure range. The number of events corresponding to this new distribution (Events') may now be calculated, applying the respective incidence rates. It can be seen that the total
events have been reduced from 126 to 105, a saving of 21 "heart
attacks" and a reduction in incidence of 17% in the whole population
of men in this age group. This mass approach can now be compared
with the "high-risk strategy". We assume that all men with systolic
pressures of 160mmHg or higher receive anti-hypertensive treatment
so that their pressures are now in the range of 130-139mmHg. The
29 events become thereby reduced to a total of 17 events, applying
the incidence rate of 105 to the 157 hypertensive men. The resulting
saving of 12 events is markedly less than the saving achieved by the
"mass strategy". In practice, the two approaches are not mututally
exclusive and their combination will lead to a greater reduction in
incidence than either approach alone.

A similar calculation may be made for serum cholesterol (Table 2)
using the same Framingham data. Shifting the distribution to the
left by 15mg/dl, from a population average of 227 to 212mg/dl, incidence would be lowered by 13%, assuming an entirely casual relationship between serum cholesterol level and disease risk. Instead
of contrasting in this case the mass strategy with the high risk
strategy, under various assumptions (the reader can carry out this
exercise by himself), let us calculate the benefit which an individual man would derive from lowering his serum cholesterol level
in different ways (Table 3), using another set of Framingham data[8].
The table shows not only how individual chances over a 10-year
period rise with rising level but how much reduction in risk might
be expected from various steps. It is apparent that the individual
risks can be lowered by around 20% through serum cholesterol reductions which are entirely within reach by relatively mild dietary
changes. The last column intends to illustrate the basis of an ob-
jection which is sometimes raised against the mass strategy. A
thousand men with levels between 220 and 234mg/dl of serum chol-

Table 1. "Mass" versus "High-Risk" strategy for primary
prevention of Coronary Heart Disease

Calculations based on 12-year follow-up of
Framingham men aged 40-54 years at entry

Shifting average systolic B.P.
from 138 to 128 mm Hg

Reducing syst. B.P. in men
with levels of ≥ 160 mm Hg

Syst.B.P.	Incidence*	N	Events	N'	Events'
<110	53	38	2	170	9
110-119	53	132	7	226	12
120-129	102	226	23	247	25
130-139	105	247	26	212	22
140-149	123	212	26	105	13
150-159	124	105	13	65	8
160-169	138	65	9	44	6
170-179	182	44	8	22	4
180-189	227	22	5	26	6
>190	269	26	7	-	-
Total	-	1.117	126	1.117	105

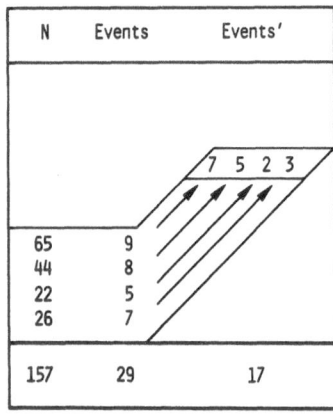

N	Events	Events'
		7 5 2 3
65	9	
44	8	
22	5	
26	7	
157	29	17

Saving: 21 events   Saving: 12 events

* incidence of myocardial infarction and cardiac death per 1,000/12 years

Source: "The Framingham Study", Section 31,1976.

Table 2. "Mass" versus "High-Risk" strategy for primary
prevention of Coronary Heart Disease

Calcualtions based on 12-year follow-up of
Framingham men aged 40-54 years at entry

Effect of shifting average serum cholesterol level from 227 mg/dl to 212 mg/dl (-7%)

Serum Cholest. mg/dl	Incidence*	N	Events	N'	Events'
<189	79	189	15	357	28
190 - 204	89	168	15	138	12
205 - 219	65	138	9	188	12
220 - 234	96	188	18	119	11
235 - 249	134	119	16	103	14
250 - 264	126	103	13	66	8
265 - 279	227	66	15	52	12
280 - 294	212	52	11	32	7
295 - 309	94	32	3	36	3
>310	222	36	8	-	-
Total	-	1.091	123	1.091	107

Lowering of incidence: $\frac{107}{123}$ = 13%

* incidence of myocardial infarction or cardiac death per 1.000/12 years

Source: "The Framingham Study", Section 31,1976.

Table 3.   Some estimates of the chances of a "Heart Attack" by serum cholesterol level and their theoretical reduction by lipid lowering in middle-aged men

Serum Cholesterol mg%	Incidence Rate* p.1000 p. year	Events p.1000 p.10 years	Individual Chances in 10 years	Cholesterol Reduction from Level:	to Level:	Reduction of 10 year chances:	Saving in Events p. 1000 men in 10 years:
1. <190	7.0	70	1:14.3				
2. 190–204	8.0	80	1:12.5	3	1	19%	16
3. 205–219	8.6	86	1:11.6				
4. 220–234	9.4	94	1:10.6	4	2	15%	14
5. 235–249	10.1	101	1:10.0	4	1	26%	24
6. 250–264	11.0	110	1: 9.1				
7. 265–279	12.0	120	1: 8.3	7	5	17%	19
8. 280–294	13.0	130	1: 7.7				
9. 295–309	14.2	142	1: 7.0	7	4	22%	26
10. 310+	18.0	180	1: 5.6				

* calculated as the yearly average of the "smoothed rates" between the ages 45–54 and 55–64 years, so as to represent the average yearly incidence rate for men between the average ages of 50 and 60 years.

"Heart Attack" is defined here by the category "coronary heart disease other than angina pectoris"

Source: The Framingham Study, 16-year Follow up, Section 26, Table 2-3-B.

esterol who lower their level to less than 190mg/dl would experience 70 instead of the original 94 events. The saving of 24 events over 10 years in a total of 1,000 men might not seem much, even though it corresponds to an average reduction in individual risk by about a quarter. It is asked whether this amount of saving is worthwhile when measured against the inconvenience, as it is called, of giving up all kinds of accustomed dietary habits. The answer is two-fold: (1) for the individual, reducing the chances of a heart attack by a quarter is far from negligible, (2) for the community, the cumulative effect of relatively small savings on the part of many people is large (as illustrated by Tables 1 and 2).

Finally, in connection with the mass strategy, mention must be made of the concept of "attributable risk". It is important to ask what proportion of the events of disease in the community are due to the presence of a given risk factor, – in other words, to what degree would incidence be reduced if no one in the population was afflicted with the risk factor. The answer provides an indication of the potential of prevention through risk factor reduction. In the field of infectious diseases, one would speak of "eradication" but the implications are different. Eradication of smallpox means disappearance of the disease altogether. It is possible to eliminate hypertension as a mass phenomenon but there will always be an appreciable number of "sporadic" cases of hypertension. Furthermore, the disease associated with hypertension, cardiovascular disease, also occurs without blood pressure elevation since hypertension is but one of the risk factors. Nevertheless, "attributable risk" is a most useful concept and will be illustrated by using hypertension

as an example (Table 4). If there were no men with systolic blood
pressures of 140mmHg and more in the population, how many fewer
heart attacks or total deaths would occur? Since systolic pressures
higher than 140mmHg constitute borderline or definite hypertension,
this value has been chosen as the cutting point. It is seen that
20% of the myocardial infarctions and cardiac deaths and 27% of the
total deaths occur at blood pressure levels above this threshold.
Clearly, even the total elimination of borderline, mild and more
severe hypertension would not eliminate more than a fifth of all
"heart attacks". However, it must be remembered that hypertension
is but one of the risk factors and the risk attributable to all the
three major risk factors is of the order of 60%. The potential of
primary prevention inherent in these figures is apparent!

Practical Approaches

The practical approaches to risk factor control in the community
must be consonant with the two main strategies. The mass strategy
requires the mobilization of all the community resources which are
relevant to the maintenance of good health. The citizens themselves
must concern themselves actively with the problem and not only make
use of the resources made available to them but insist that any
facilities which are lacking be provided through administrative and
individual action. The medical services should have preventative
as well as curative components. There must be a joint effort on the
part of the different health agencies and the medical profession,
the social groups and organizations in the community and those ad-
ministratively and politically responsible for the population of the
area. Opinion leaders play an important role. The allied health
professions must be available to provide special services, particularly in the field of health education and motivation. In the long
range, the task is not so much to lower risk factors which have been
allowed to develop but to prevent their occurrence from youth on.
Therefore, the family and the schools have to be involved very deeply
in any community prevention program. Provisions have to be made to
detect and treat persons at high risk. The links between curative
and preventive medicine will depend largely on the practicing physicians in the community, according to the prevailing structure of
medical practice.

There are no hard and fast or universally valid rules for creating or strengthening preventive health services. Much depends on
the facilities which already exist and every effort should be made
to build up from any such structures. Most if not all of the building blocks of a preventive community control system exist in the
model of the North Karelia Study in Finland and the report of this
project may be consulted for guidance on the principles involved[9].

Table 4.   Excess events of myocardial infarction (MI) or cardiac
           death (CD) and all deaths attributable to systolic blood
           pressure >140mmHg

           Framingham Men, aged 40-54 at entry, 12-year follow-up

Syst.B.P.	Percentage Distribution	Actual Events in 12 years		Expected Events in 12 years		Excess Events attributable to Syst.BP >140 mm Hg	
		MI or CD	All deaths	MI or CD	All deaths	MI or CD	All deaths
<140	58	52.0	48.7	52.0	48.7	-	-
140-149	28	34.6	34.4	25.2	23.5	9.4	10.9
160-179	10	15.6	17.5	9.0	8.4	6.6	9.1
≥180	4	9.9	14.7	3.6	3.4	6.3	11.3
Total	100	112.1	115.3	89.8	84.0	22.3	31.3

Calculated from "The Framingham Study", Section 31 (April 1976).

Excess events due to elevated systolic pressure: MI or CD $\frac{22.3}{112.1}$ = 20%

All deaths $\frac{31.3}{115.3}$ = 27%

People are individuals but they also belong to their families,
social groups and the community in which they live. The "community
spirit" is important but communities need the support of the country
of which they are a part. Community medicine must come from within
the community but it is also essential that broad national health
policies and health legislation be consonant with local aims and
activities. If the promotion of prevention becomes part of a national
health policy, it stands a better chance of succeeding on the community level.

Summary

Risk factor control in the community is a joint effort which
involves collaboration between the people whose health is to be pre-
served and the medical, social and administrative organizations and
agencies who are responsible for or concerned with health maintenance.
Evidence is presented that risk factors can be reduced in the community and that the lowering of risk factors, on the basis of all
the available evidence, reduces the risk of coronary heart disease.
The long-range goal is to prevent the development of risk factors
rather than their treatment when they are already too high. Risk
factor control should start early in life and is thus a family matter

and a concern of the schools. Preventive cardiology on the community
level includes not only primary prevention but also secondary prevention for patients with coronary heart disease. Primary prevention
must pay special attention to persons at high risk (the "high-risk
strategy") but it is shown that effective risk factor control in the
community requires a "mass strategy" which aims to motivate the entire population to adopt life styles which favor low levels of risk
factors. In this fashion, the whole distribution of risk factors in
the population will be shifted to lower levels. The mass – and high-
risk strategies are complementary. The first strategy is a necessity
for the adequate lowering of the burden of disease in the community,
the second approach to those at high-risk constitutes an ethical
responsibility toward providing good preventive medical care. Effective and efficient preventive programs require a high degree of
motivation to follow sensible living habits on the part of each person but these individual efforts must be constantly reinforced and
strengthened through community participation and action, as well as
political and legislative support on the national level.

REFERENCES

1. F.H. Epstein, Primary Prevention, in: "Comprehensive Coronary
   Care," H. Denolin, H.J.C. Swan, and Z. Pisa, eds., Marcel Dekker
   Inc., New York (in press).
2. Secondary Prevention in Myocardial Infarction Survivors, Heart-
   beat, International Soc. and Federation of Cardiology, 3:1 (1980)
3. G.S. May, K.A. Eberlein, C.D. Furbery, E.R. Passamani, and D.L.
   DeMets, Secondary prevention after myocardial infarction, Progr.
   Cardiovasc.Dis. 24:331 (1982).
4. F.H. Epstein, Coronary heart disease – geographical differences
   and time trends, in: "Atherosclerosis VI," Springer Verlag,
   Berlin, Heidelberg, New York (in press).
5. F.H. Epstein, Coronary heart disease epidemiology revisited:
   clinical and community aspects, Circulation 48:185 (1973).
6. G. Rose, Strategy of prevention: lessons from cardiovascular
   disease, Brit.Med.J. 1:1847 (1981).
7. The Framingham Study, an epidemiological Investigation of
   cardiovascular disease, W.B. Kannel and T. Gordon, eds., Section
   31, The Results of the Framingham Study applied to four other
   US-based studies of cardiovascular disease, U.S. Dept. Health,
   Education and Welfare, PHS, NIH, DHEW Publ.No (NIH) 76-1083
   (1976).
8. The Framingham Study, An epidemiological investigation of
   cardiovascular disease, Section 26, Some characteristics related
   to the incidence of cardiovascular disease and death – Framingham
   Study, 16-year follow-up. US Govt. Printing Office, Washington
   D.C. (1970).
9. North Karelia Project: Community Control of Cardiovascular
   Diseases, Regional Office for Europe, World Health Organization,
   Copenhagen (1981).

PREVENTIVE APPROACH TO CORONARY HEART DISEASE IN PHYSICIAN'S PRACTICE, HOSPITAL MEDICINE, OCCUPATIONAL MEDICINE, AND COMMUNITY HEALTH CARE

Kalevi Pyörälä

Department of Medicine
Univeristy of Kuopio, Kuopio, Finland

Physicians have traditionally been trained for diagnosing
diseases and treating patients - people who already are sick. Until
now, preventive approach to diseases has not formed any large part
of medical practice in most branches of medicine. In the cardiovascular field, despite increasing knowledge concerning possibilities
for the prevention of most of the major cardiovascular diseases,
practical advances have been slow. The following review will deal
with a preventive approach to coronary heart disease (CHD) - a mass
health problem which evidently is preventable to a large extent.
The background of the prevention of CHD will be briefly summarized.
After reviewing the strategies for the prevention of CHD, according
to the principles outlined recently by the WHO Expert Committee on
the Prevention of Coronary Heart Disease[1], the central elements of
a preventive approach to CHD in different settings of health care -
in physician's practice, hospital medicine, occupational medicine,
and community health care - will be discussed. A particular emphasis
will be given to the role of physicians in the prevention of CHD.
The recent Bethesda Conference report on the Prevention of Coronary
Heart Disease[2] is an outstanding and detailed review on this subject
focusing especially on the role of physicians. Although the recommendations of the Bethesda Conference are specifically adapted to
the circumstances of the United States, most of them are generally
applicable.

BACKGROUND FOR THE PREVENTION OF CORONARY HEART DISEASE

Coronary atherosclerosis which leads to clinical manifestations
of CHD in middle-age or later starts to develop already in adolescence or early adult life. Sudden death is not an uncommon first

manifestation of CHD and in those people in whom CHD manifests itself as an acute myocardial infarction or angina pectoris, coronary atherosclerosis is already extensive. Although medical and surgical advances have improved the prognosis of individual patients with symptomatic CHD, the possibilities of these treatments to reduce morbidity and mortality at population level are relatively small.

In addition to the insidious nature of CHD, the large size of the CHD problem from the public health point of view is an important argument for a preventive approach. In many developed countries cardiovascular diseases cause one half of the total mortality and among cardiovascular diseases CHD causes the largest number of deaths. Much of the CHD mortality occurs among working-age population and even larger numbers of people become prematurely disabled due to CHD.

The well known large differences in the CHD mortality rates between populations and changing trends within populations during relatively short periods of time indicate that the population rates for CHD must be strongly influenced by environmental or life-style factors. The example of Japan and some Southern European countries still showing low or relatively low CHD rates despite advanced socio-economic development indicates that CHD as a mass public health problem is not an inevitable concomitant of socio-economic development, but is apparently more connected to a "western" type of life-style and health behavior. Thus, it is evident that there is a great potential for the prevention of CHD.

A large body of scientific knowledge has accumulated concerning the causes of atherosclerosis and CHD. Although there still remains much space for further research, the WHO Expert Committee on the Prevention of Coronary Heart Disease[1] concluded: "It is the judgement of the Committee that major determinants of population death rates of CHD have now been identified: an inappropriate national diet aggravated by physical inactivity and overweight (reflected in mass raising of blood lipids and blood pressure), and widespread cigarette smoking".

Prevention programs carried out in whole communities, like the Stanford Three Community Study[3] and the North Karelia Project[4] have shown that it is feasible to lower CHD risk factor levels in populations by health education measures. Large controlled trials have demonstrated that a lowering of elevated blood pressure by drug treatment results in a reduction of cardiovascular mortality, including favorable trends in CHD mortality[5,6]. A good feasibility of implementation of an effective hypertension control program at community level has been demonstrated in connection with the North Karelia Project[4]. A recent study from Oslo has shown a reduction of CHD mortality in middle-aged high-risk men following a long-term lowering of plasma cholesterol by dietary fat intake modification

and reduction of cigarette smoking[7]. Several major cardiovascular
risk intervention trials using a multifactorial design and aiming
to measure the effect of risk reduction on CHD mortality are now
in progress in the United States, in different European countries
and in the USSR. These trials are, however, being carried out among
middle-aged people and most of them among people with high risk.
Thus, these trials are testing the effects of risk reduction in
people many of whom already have an extensive coronary atherosclerosis. For many reasons it will never be possible to carry out
large-scale and life-time scientific experiments on primary prevention of CHD in whole populations. Therefore, decisions concerning
the primary prevention of CHD at population level have to be made
without such a final field experiment. The ongoing large cardiovascular risk intervention trials will probably give valuable
additional information in this respect, but even without results of
these trials the WHO Expert Committee on the Prevention of Coronary
Heart Disease[1] concluded with respect to the need for changes in
key areas of life-styles important for CHD prevention in whole
populations that "the balance of evidence indicates a sufficient
assurance of safety, and a sufficient probability of major benefits,
to warrant action now". Although the interpretation of "natural
experiments" in countries showing markedly declining CHD mortality
trends is complex, data from Australia[8], United States[9] and Finland[10]
show that in these countries declining trends of CHD mortality were
preceded and paralleled by favorable changes in life-styles and
major risk factors - diet, cigarette smoking and hypertension.
These "natural experiments" are very encouraging and give further
support to the view that the preventive approach to CHD is going
in the right direction.

STRATEGIES FOR THE PREVENTION OF CORONARY HEART DISEASE

     A comprehensive strategy for the prevention of CHD includes
three components[1]:

     1. A population strategy for altering the mass characteristics
of life-style and environment and their social and economical determinants, which are the underlying causes of mass CHD.

     2. A high-risk strategy bringing preventive care to indi-
viduals at high risk.

     3. Secondary prevention aimed against recurrences and pro-
gression of disease in those already afflicted.

     The high-risk strategy and secondary prevention are of impor-
tance for a small segment of the population. However, in countries
with a high incidence of CHD the majority of people have elevated
levels of risk factors, often involving more than one risk factor,

Most of the CHD events occur among these people with moderate
elevations of risk factor levels. Therefore, to bring help to the
majority of people at risk, the preventive approach has to be directed to the whole population.

In a few developed countries, like Japan and some Southern
European countries, and in most developing countries, life-style
and associated CHD risk factors have not yet assumed a pattern
leading to high population rates of CHD. In these countries it will
be urgent to try to maintain the current favorable status. The term
"primordial" prevention has been coined for the primary prevention
of CHD in such populations in which CHD still is virtually non-
existent.

A population approach to CHD prevention has the following
intermediate objectives in terms of life-style and risk factor modification:

1. Lowering of the population distribution of plasma cholesterol.

2. Lowering of the population distribution of blood pressure.

3. Getting non-smoking as a predominant behavior.

4. Avoidance of sedentary life and control of obesity.

The WHO Expert Committee on the Prevention of Coronary Heart
Disease[1] defined a population average value for plasma cholesterol
of under 200mg/dl (5.2mmol/1) as a target value, since such a plasma
cholesterol average value would be associated with a relatively low
frequency of CHD. To achieve this in populations with much higher
plasma cholesterol average values will involve progressive changes
of the dietary pattern including a reduction in saturated fats to
less than 10% of total calories and dietary cholesterol intake to
under 300mg average for adult individual and an increase in complex
carbohydrate consumption. The composition of some national diets
requires that some of the reduction in saturated fats has to be
made up by mono- and poly-unsaturated fats. As to the controversial
question of recommendable poly-unsaturated fat intake, there is both
observational and experimental data to suggest that an appropriate
upper limit could be 10% of total calories as poly-unsaturated fats.
Lowering the total fat intake is important in countries in which
the level of physical activity is low and obesity common. In such
countries it would be appropriate to limit the total fat intake to
25-30%. In attempts to achieve optimal plasma cholesterol levels,
prevention or correction of obesity is apparently of greater importance than recognized so far. Therefore, through their effects
on obesity, a moderation in total calorie intake and increased
leisure-time activity are important in attempts to achieve optimal
population average levels for plasma lipids.

An effective program for drug therapy of sustained high blood pressure is essential for the prevention of CHD. Further development of non-pharmacological, hygienic measures for controlling elevated blood pressure and, hopefully even preventing it, is however one of the key areas of research now. Rose[11] has calculated that all the reduction of cardiovascular mortality by current antihypertensive treatment might be equalled, if it were possible to achieve a downward shift of the whole blood pressure distribution in the population by 2-3mmHg by hygienic measures. This might be possible through a gradual reduction of the population consumption of salt towards an average of 5 grams or less per adult individual. Control of obesity is also of central importance in the hygienic control of elevated blood pressure and probably also in its prevention. In addition, avoidance of excessive alcohol consumption has proved to be important in this respect.

There is no longer any argument about non-smoking as a behavior which should be predominant in the population. Regular physical activity is important in the prevention and control of obesity and through this, but also otherwise, it will help in the lowering of CHD risk factors.

As already mentioned, coronary atherosclerosis starts to develop in adolescence or in early adult life, along with the appearance of the major risk characteristics or precursors of CHD, elevated plasma cholesterol and blood pressure, and the beginning of smoking. Lifestyles behind these precursors of CHD are adopted early in life and therefore it is important that a population approach for the prevention of CHD includes effective strategies aiming to prevent the development of these precursors of CHD in youth. This calls for an approach through families and schools.

PREVENTIVE APPROACH TO CORONARY HEART DISEASE IN DIFFERENT
SETTINGS OF HEALTH CARE

The preventive approach to CHD in different settings of health care will be discussed in the following with an emphasis on the role of physicians, but participation of other health personnel and involvement of non-medical sectors will also be briefly mentioned in order to give perspective to the role of physician in the prevention of CHD in the community.

Physician's practice

Although physicians working in first-line medical practice with individual patients are becoming increasingly aware of the possibilities and needs of a preventive approach to CHD, the concept of prevention as an integral part of a good medical practice should

be promoted in order to motivate those physicians who have so far
not been actively applying these principles in their work. Professional and scientific associations of physicians and medical
educators responsible for both undergraduate and postgraduate
teaching have an important role in the promotion of this concept.

Secondary prevention - attempts to prevent recurrent heart
attacks or progression of the disease in patients who have survived
a myocardial infarction or otherwise have an established CHD - is
an area of preventive approach with which the physicians are best
familiar. Important advances have recently been made in the medical
and surgical treatment of patients with symptomatic CHD and much
attention to these forms of treatment has tended to overshadow the
potential importance of measures aiming to reduce risk factor levels
in order to prevent the progression of underlying coronary atherosclerosis. Although the extent of myocardial damage is the main
determinant of the short-term prognosis after an acute myocardial
infarction, the progression of coronary atherosclerosis with super-
imposed thrombotic phenomena becomes a decisive prognostic factor
in long-term survivors and in patients with uncomplicated angina
pectoris. Therefore, preventive measures aiming to reduce risk
factor levels as appropriate for each individual patient should be
an integral part of a comprehensive care of patients with diagnosed
CHD. The Councils on Atherosclerosis, Epidemiology and Prevention,
and Rehabilitation of the International Society and Federation of
Cardiology have recently published joint recommendations on the
secondary prevention of CHD, with a particular emphasis on risk
factor modification[12].

High-risk strategy - institution of preventive measures in the
care of persons identified as high-risk individuals with respect to
future risk of CHD - is getting increasing importance in medical
practice. Increasing awareness of CHD and its risk factors among
the public leads to an increasing number of requests for CHD risk
assessment, particularly among persons whose relatives have died
of CHD or experienced other CHD events at a relatively young age.
CHD risk assessment is also becoming a part of regular "health
check-ups" or entrance medical examinations of some professional
groups. Health insurance examinations also give opportunities for
cardiovascular risk assessment. In fact, CHD risk assessment can
be included with relatively small extra expense as a part of any
medical examination. In order to be able to apply preventive
approach to the high-risk individuals, each physician needs a plan
for their identification and management. An expert group of the
European Society of Cardiology has in 1978 prepared a guide booklet
for the practising physicians on the prevention of CHD[13] and the
principles outlined in that booklet still remain valid.

The assessment of CHD risk includes, in addition to the family
history, medical history and physical examination, information

about smoking habits, eating pattern and physical activity, the measurement of height and weight, blood pressure measurement and determination of plasma total cholesterol. With the exception of the plasma cholesterol determination, all this information can be collected at a single visit to the physician's office and auxiliary staff can be trained to help in the collection of information on smoking, eating and physical activity. Additional tests, like an electrocardiogram, determination of fasting blood glucose or assessment of glucose tolerance, or determination of plasma lipid fractions may be needed, depending on the age of the subject, family history, medical history, findings at physical examination and plasma total cholesterol level.

Assessment of individual CHD risk factor status gives a good basis for specific advice with respect to those areas of health behavior in which modification is needed. It is important that the physician takes his or her time to explain to the patient what factors should be modified, why this is necessary and how these changes can be accomplished. In attempts to stop smoking, in the improvement of eating pattern, in the correction of obesity or in the increase of physical activity, it is useful to set distinct goals and if several health behavior areas need modification, it may be better not to attempt to modify them all immediately. In order to accomplish permanent changes in health behavior it is essential that the patient becomes motivated to make his or her own decisions to make appropriate changes. The physician has to provide advice, help and support in changing behaviors related to the risk. To achieve and maintain changes in health behavior it is essential to arrange a systematic follow-up and adjust further advice according to the results achieved. Since training in health education has not belonged to the traditional medical education, possibilities to get training in this field have to be arranged for physicians and health team members working with them. Available health education resources in the community provide often a useful support system for health education started at a physician's office.

The interest and enthusiasm of the physician is the key element in the implementation of a preventive approach to CHD as a part of everyday practice. Needless to say, physicians should set a good example through their own health behavior.

Hospital medicine

Hospital medicine is traditionally involved in the care of patients with acute myocardial infarction and in the diagnostic evaluation and medical and surgical treatment of symptomatic CHD. Assessment of CHD risk factor status and appropriate advice concerning risk-related health behavior modification should be an integral part of a comprehensive long-term care of survivors of

myocardial infarction and other patients with established CHD.
It is important that hospitals and their out-patient clinics try
to ensure the continuation of patient counselling, when the patient
is referred back to his or her own physician or to the primary
health care centre of the community.

The hospitals have in most countries been rather uninterested
in providing health education which goes outside their task of care
of those patients with specific diseases. Yet the hospitals have
a great potential to act as models in some key areas of health
behavior related to CHD prevention, particularly in the prudent
eating pattern using hospital diet as a demonstration channel and
in the promotion of non-smoking through hospital's own measures to
favor non-smoking behavior. Even broader health education programs
with respect to CHD prevention can be developed at hospitals as
joint efforts of physicians, nurses, dietitians, psychologists and
other personnel. Such programs would not only improve the comprehensive care of post-myocardial infarction patients and other
patients with established CHD, but suitable elements of such programs
could be used in contacts with non-cardiovascular patients in
counselling them with respect to CHD prevention.

The number of hospital staff and employees is large and they
should not be forgotten in the development and implementation of
health education programs at hospitals. Through their personal
interest and involvement they will be better able to apply the preventive approach to their patients.

The contact and collaboration of hospitals with prevention
programs in the community is important. Hospital staff should help
community prevention programs as consultants and teachers, as appropriate in local circumstances.

Occupational medicine

Large segments of adult population can be reached through
occupational medicine and in many instances it is possible to
approach also the families of employees. Work places with large
numbers of employees often form a suitable setting for different
elements of CHD prevention. Hypertension detection and control
programs can be successfully arranged in industry. Work place gives
a good social setting for campaigns and group activities for smoking
cessation, control of obesity, improved nutrition, and increased
exercise. Useful experience from the application of mass health
education and individual counselling in factories is accumulating
from the WHO European Collaborative Group Heart Diseases Prevention
Project[14-16].

Community health care

Many essential elements of the population strategy for CHD prevention go far beyond the traditional health care system. It is, however, of great importance to the success of CHD prevention programs in the community that physicians and other health personnel working in primary health care, family medicine and school medicine are actively involved in these programs. In the Finnish demonstration program of cardiovascular disease prevention in the community, the North Karelia Project[4], the whole community health care system was engaged in its activities and this proved to have a decisive influence in the activation of the whole community in health promotion.

Although the prevention of CHD at the level of the whole population extends outside the conventional medical practice, it has to be recognized that the individual approach - primary prevention in high-risk individuals and secondary prevention in patients with established CHD - and the population approach are supplementary to each other. Primary health care physicians have, in collaboration with physicians working in local hospitals, an important role in ensuring the continuity of preventive care of patients who have survived a myocardial infarction or in whom CHD has otherwise been diagnosed. As to the preventive care of high-risk individuals, the principles outlined above for physician's practice are basically similar for primary health care and specialized care levels. Physicians working in primary health care practice or in community health centers should, together with their colleagues and other staff, make efforts in developing health education and promotion activities as a part of their everyday work and they should also, through their leadership and participation as consultants and teachers, stimulate the development of other health education and promotion activities in the community.

The development of an effective hypertension detection and treatment program can well be established at the community health care level, as demonstrated by the North Karelia Project[4]. The Stanford Three Community Study[3] and the North Karelia Project[4] have shown that improved eating and smoking patterns can be achieved through effective health education programs using mass media supported by individual and group health education and simultaneous activation of the community resources towards changes in these health behavior areas through other means. Further experience is currently being obtained from many second-generation demonstration projects in several countries.

Although much remains to be learned through research and practical experience concerning effective strategies for CHD prevention at the level of whole populations, the time has come to promote vigorously national public health policies for the prevention of

CHD. The response of national health administrators to this challenge is important and for the improvement of smoking and eating
patterns commercial, industrial and agricultural circles have to be
involved. Legislative measures may be necessary in certain areas,
particularly in the promotion of non-smoking. The creation of
better possibilities for regular physical activity has influences
on town planning and architecture.

National cardiological societies and other scientific and
medical organizations, together with voluntary layman organizations
in the cardiovascular field, should work in liaison with the govern-
mental health administrators for the development of effective national
strategies for the prevention of CHD involving, in addition to
medical profession and other health personnel and the whole health
care system, also community organizations, schools, work places, as
well as mass communications.

REFERENCES

1. WHO Technical report series, No 678, Prevention of coronary
        heart disease, report of a WHO Expert Committee, WHO, Geneva,
        (1982).
2. Eleventh Bethesda Conference, Prevention of Coronary Heart
        Disease, Amer. J. Cardiol., 47:713 (1981).
3. J. W. Farquhar, N. Maccoby, P. D. Wood, J. K. Alexander, H.
        Breirrose, B. W. Brown, Jr., W. L. Haskell, A. L. McAlister,
        A. J. Meyer, J. C. Nash and M. P. Stern, Community education
        for cardiovascular health, Lancet, 1:1192 (1977).
4. P. Puska, J. Tuomilehto, J. Salonen, A. Nissinen, J. Virtamo,
        S. Björkqvist, K. Koskela, L. Neittaanmäki, T. Takalo, T. E.
        Kottke, J. Mäki, P. Sipilä and P. Varvikko, Community Control
        of Cardiovascular Diseases: The North Karelia Project, WHO,
        Regional Office for Europe, Copenhagen (1981).
5. Hypertension Detection and Follow-up Program Cooperative Group,
        Five-year findings of the Hypertension Detection and Follow-up
        Program. I: Reduction in mortality of persons with high
        blood pressure, including mild hypertension, J. Amer. Med.
        Ass., 244:2562 (1979).
6. Report by the Management Committee, The Australian therapeutic
        trial in mild hypertension, Lancet, 1:1261 (1980).
7. I. Hjermann, K. V. Byre, I. Holme and P. Leren, Effect of diet
        and smoking intervention on the incidence of coronary heart
        disease, Lancet, 2:1303 (1981).
8. T. Dwyer and B. S. Hetzel, A comparison of trends of coronary
        heart disease mortality in Australia, USA and England and
        Wales with reference to three major risk factors - hyper-
        tension, cigarette smoking and diet, Int. J. Epidemiol.,
        9:65 (1980).
9. J. Stamler, Primary prevention of coronary heart disease: the
        last 20 years, Amer. J. Cardiol., 47:722 (1981).

10. K. Pyörälä and T. Valkonen, The high ischaemic heart disease mortality in Finland. International comparisons, regional differences, trends and possible causes, In: "Medical Aspects of Mortality Statistics", H. Boström and N. Ljungstedt, ed., Almquist & Wiksell International, Stockholm (1981).
11. G. Rose, Strategy of prevention: lessons from cardiovascular disease, Br. Med. J., 282:1847 (1981).
12. Secondary prevention in myocardial infarction survivors. Joint recommendations by the ISFC Scientific Councils on Arterio-sclerosis: Epidemiology and Prevention, and Rehabilitation, Heart Beat No. 3:1 (1980).
13. European Society of Cardiology, "Preventing Coronary Heart Disease: A guide for the Practising Physicians", Van Gorcum & Comp. B. V., Assen (1978).
14. G. Rose, R. F. Heller, H. Tunstall Pedoe and D. G. S. Christie, Heart disease prevention project: a randomized controlled trial in industry, Br. Med. J., 1:747 (1980).
15. M. Kornitzer, G. De Backer, M. Dramaix and C. Thilly, The Belgian Heart Disease Prevention Project: Modification of the coronary risk profile in an industrial population, Circulation 61:8 (1980).
16. M. Kornitzer, M. Dramaix, F. Kittel and G. De Backer, The Belgian Heart Disease Prevention Project: Changes in smoking habits after two years of intervention, Prev. Med., 9:496 (1980).

MULTIVARIATE PREDICTION OF CORONARY HEART DISEASE

Gino Farchi

Laboratorio di Epidemiologia e Biostatistica
Istituto Superiore di Sanita
Roma, Italia

INTRODUCTION

One of the major achievements of the past 30 years' cardio-vascular epidemiology has been the identification of the so called coronary risk factors and the possibility to assess their power in prediction of coronary events.

Risk factors are those individual characteristics which, if measured in healthy people, are able to identify subgroups of population which, in a defined period of time, will experience the oc-currence of an excess of disease if compared with other subgroups of the same population.

It is useful to study the relationship between each single factor and the manifestation of the disease and indeed many uni-factorial studies have been conducted all over the world. It is likely more useful to study the predictability of a whole set of risk factors having in mind the underlying hypothesis that the whole set of risk factors will predict the disease better than each risk factor considered by itself.

Many studies have shown that this hypothesis holds; for example, it is reported[1] that Q_5/Q_1 (ratio between the number of cases of CHD in fifth quintile and the number in the first quintile) is 2.60 for systolic blood pressure, 3.38 for serum cholesterol, 6.50 for blood pressure and cholesterol used all together in a bivariate way, 8.67 adding other factors in a multivariate way. It has also been demon-strated[2] that increasing more and more the number of risk factors in a multivariate model, a saturation effect is observed, i.e., by adding new factors to a set of major risk factors a new set is ob-tained which is not better predictive than the old one.

In longitudinal multifactorial studies in which many risk factors are measured, the need for multivariate predictive models to analyze data and interpret the results arises.

The following main points related to the multivariate prediction of coronary heart disease, will be discussed:

(a) Linear and logistic models,
(b) Statistics and statistical tests,
(c) Applicability of a model to populations different from those in which the set of coefficients of the model was estimated.

LINEAR AND LOGISTIC MODELS

Historically, the traditional analytic method of epidemiologists was the multiple cross-classification, but such method quickly becomes impracticable as the number of variables to be investigated does increase. Thus if 10 variables are under consideration and each variable, split for example in three levels, has to be studied, there would be 59,049 cells in the multiple cross-correlation. Even with only 10 subjects for each cell, a cohort of about 600,000 people would be required. A mathematical model was necessary for the treatment of many interacting variables in a more efficient way.

The most simple way for adding the effect of many factors in the estimation of a value related to the global risk is to do that by means of a linear model, i.e., using a linear regression equation:

$$y = a + b_1 \cdot x_1 + b_2 \cdot x_2 + \ldots + b_k \cdot x_k$$

where y is the dependent variable (status after t years in our case; 0 if non event and 1 if event);

$x_1, x_2 \ldots x_k$ are the independent variables (risk factors);

$b_1, b_2 \ldots b_k$ are the regression coefficients;

a is the intercept;

k is the number of factors.

One disadvantage of this kind of model is however that y has no probabilistic meaning, being a not-bounded function which can also assume negative values. This consideration, among other things, led Cornfield[3] to the suggestion of a different model. He considered the case of k variables, say $x_1 \ldots x_k$, and assumed that the multivariate frequency distributions of those who would and those who would not develop the disease could be represented by two known mathematical functions, say $f_1(x_1 \ldots x_k)$ and $f_0(x_1 \ldots x_k)$. In that case the probability $P(x_1 \ldots x_k)$ that an individual characterized by the variable values $x_1 \ldots x_k$, would develop the disease is given by the Bayes' formula

$$P(x_1 \ldots x_k) = \cfrac{1}{1 + \cfrac{1-p}{p} \cdot \cfrac{f_0(x1 \ldots x_k)}{f_1(x_1 \ldots x_k)}}$$

where p is the unconditional probability of developing the disease. In particular, if the frequency distributions, f_0 and f_1, are multivariate normal, with the same variances and covariances

$$P = \cfrac{1}{1 + \exp(a + b_1 \cdot x_1 + b_2 \cdot x_2 + \ldots + b_k \cdot x_k)}$$

where b_i are the coefficients of the model to be estimated by means of the discriminant function theory or by a least square approach suggested by Walker-Duncan[4].

This function is referred as the Multiple Logistic Function (MLF). It is bounded between 0 and 1 and has probabilistic meaning, under the specified conditions. The main limitation to its use is due to the assumption of multivariate normality which is rarely satisfied, as the same Cornfield points out.

STATISTICS AND STATISTICAL TESTS

Statistics and statistical tests to judge the "goodness" of a solution (a model whose coefficients were estimated from experimental data) are needed in order to measure:

1. the statistical significance of the solution;
2. the significance of the estimate of each single coefficient and therefore the contribute of the risk factor associated to that coefficient;
3. the discrimination between cases and non cases that is the steepness of risk gradient.

Several criteria may be considered for determining how well the estimated regression of CHD incidence conditional on risk factors fits the experimental data. One may be the usual chi-square statistic used to compare the observed and expected number of cases in each decile class of risk; another method may be based on the explained proportion of variance[1]. The most frequently used is the Likelihood Ratio Statistic (LRS) that allows to assess whether the data are significantly better fitted by taking the risk factors into account

than by assuming everyone to have the same risk. Kendall[5] shows
that as n (number of observations) increases, the LRS tends to a
chi-square distribution with k degrees of freedom, where k is the
number of risk factors. It can be seen that the value of LRS depends
on the agreement between observed and expected status (case or non-
case) and also on the number of subjects and the number of cases in
the sample. Woodbury et al.[6], discussed deeply the properties of
logistic multiple regression especially in order to face problems
arising in comparing longitudinal studies with different length.

The statistical significance of the single coefficient is evaluated by the t-value, that is the ratio of the coefficient to its
standard error. The common confidence level, $p \leqslant 0.05$, equivalent
to $t \geqslant 1.96$, is normally used to reject the null hypothesis that the
coefficient is 0 and hence that there is no significant relationship
between factor and incidence.

There are several statistics used to measure the discrimination
between "cases" and "non cases" obtained with a solution of the MLF,
that is how well, for a particular estimate of the coefficients, the
observed cases are classified in higher deciles of risk rather than
in the lower ones. The relative risk statistics are very simple and
are given by the ratio of the number of cases in the upper percentile
over the number in the lowest percentile (D_{10}/D_1 if deciles are used,
Q_5/Q_1 if quintiles). A criticism is that these statistics take in
account only the first and the last percentil disregarding the number of cases classified in between.

To avoid this inconvenience Menotti et al.[2] proposed the use of
the location of the center of gravity of the distribution of cases
in decile of risk, that is

$$M = \sum_{i=1}^{10} c_i \cdot i / N_c$$

where c_i is the number of cases classified in the i-th decile. To
clarify the meaning of M, let's note that when a case moves from
decile i to decile i + 1, the value of M increases by $1/N_c$ where N_c
is the total number of cases.

CROSSING-OVER SOLUTIONS BETWEEN DIFFERENT POPULATIONS

What we discussed so far is description rather than prediction
since we gave criteria to measure how well data on risk factors and
incidence fit into the model. One of the main problems in the evaluation of the predicting power of such models concerns their applicability to populations different from those which produced the set of

coefficients of the multivariate function. If a solution is applied
back to the population (i.e. to single individuals of the population)
which produced the solution, a probability estimate of the individual
risk is obtained and with few major factors a rather good classification is reached (typically about 30%-40% of cases in the upper
decile of risk).

When the same operation is made by applying the risk function
to the individuals of another population, the usual outcomes are the
following: a) the correlation between observed and predicted, now
we can say prediction, cases is equally good; b) the total number of
the predicted cases does not correspond to the total number of the
observed cases, indicating that the absolute risk is not accurately
estimated, since time by time the incidence can be over or under-
estimated. Examples of such incongruencies are those provided by
crossing solutions between the population samples of the Seven
Countries Study, even exploiting the advantage that they had been
studied with the same methodology. When employing the 5-year follow-
up data, the solution produced from the American samples provided
an over-estimation of incidence of about 50% when applied to the
European samples[7]. Crossing solutions based on 10-year follow-up
data between Northern and Southern Europe, wrong estimations were
still present whereas a more precise estimate was obtained when
crossing the Northern European with the American samples giving an
error of only 10%.

Likewise the Seven Countries Study, similar results are obtained
in the Framingham study whose model for white middle-aged men over-
estimates twice the incidence in Puerto Rico and in Honolulu Japanese
men and three times the observed incidence in Yuguslav urban population[8,9].

On the contrary, good ranking and good prediction of incidence
are obtained when the crossing is made between populations characterized by similar cultures. Framingham risk function works very
well if applied to other samples of middle-aged white men of the
United States[10]. An Italian 6-factor risk function whose coefficients
were estimated from the Italian cohorts of the Seven Countries Study,
shows a very good predictability when applied to the Italian factories
of the W.H.O. multifactorial preventive trial on Ischaemic Heart
Disease[11]. It is worth noting that the risk function was based on
a population studied more than 10 years before the W.H.O. trial.

Hence the ranking of CHD risk may be confidently extended to
other samples whereas the prediction of incidence can only be attempt-
ed employing risk functions to populations of similar countries or,
better, of the same country. The good predictivity of the Italian
solution has given the hints for the production of a Manual of Coronary Risk[12] for Italian national use.

REFERENCES

1. T. Gordon, W.B. Kannel, and M. Halperin, Predictability of cor-
    onary heart disease, J.Chron.Dis. 32:427 (1979).
2. A. Menotti, R. Capocaccia, S. Conti, G. Farchi, S. Mariotti,
    A. Verdecchia, A. Keys, M.J. Karvonen, and S. Punsar, Identify-
    ing subsets of major risk factors in multivariate estimation
    of coronary risk, J.Chron.Dis. 30:557 (1977).
3. J. Cornfield, Joint dependence of coronary heart disease on
    serum cholesterol and systolic blood pressure: a discriminant
    function analysis, Fed.Proc. 21, Suppl. 11:58 (1962).
4. S. Walker and D.B. Duncan, Estimation of probability of an
    event as a function of several independent variables, Biometrika
    54:167 (1967).
5. M.G. Kendall and A. Stuart, "The advanced theory of statistics,"
    Griffith, London (1963).
6. M.A. Woodbury, K.G. Manton, and E. Stallard, Longitudinal models
    for chronic disease risk: an evaluation of logistic multiple
    regression and alternatives, Int.J.Epid. 10:187 (1981).
7. A. Keys, C. Aravanis, H. Blackburn, F.S.P. Von Buchem, R. Buzina,
    B.S. Djordjevic, F. Fidanza, M.J. Karvonen, A. Menotti, V. Puddu,
    and H. Taylor, Probability of middle-aged men developing coronary
    heart disease in five years, Circulation 45:815 (1972).
8. T. Gordon, M.R. Garcia-Palmieri, A. Kagan, W.B. Kannel, and J.
    Schiffman, Differences in coronary heart disease in Framingham,
    Honolulu and Puerto Rico, J.Chron.Dis. 27:329 (1974).
9. D. Kozarevic, B. Pirc, Z. Racic, T.R. Dawber, T. Gordon, and W.
    J. Zukel, The Yugoslavia cardiovascular disease study - 2.
    Factors in the incidence of coronary heart disease, Amer.J.Epid.
    104:133 (1976).
10. D. McGee and T. Gordon, The results of the Framingham study
    applied to four other U.S.-based epidemiologic studies of
    cardiovascular disease, in: "The Framingham Study," W.B. Kannel
    and T. Gordon, eds., DHEW (NIH) 76-1083, Washington D.C. (1976).
11. A. Menotti, R. Capocaccia, G. Farchi, S. Mariotti, and A.
    Verdecchia, The estimation of coronary heart disease incidence
    in a population from the risk function of another, Przeg.Lek.
    38:761 (1981).
12. R. Capocaccia, G. Farchi, S. Mariotti, A. Menotti, and A.
    Verdecchia, La previsione della cardiopatia coronarica a partire
    dai valori individuali dei fattori di rischio. Metodologia per
    la messa a punto di un manuale del rischio coronarico, ISS
    Reports ISSN-0391-1675, 15:1 (1980).

SCREENING FOR EARLY DETECTION OF ASYMPTOMATIC CORONARY ARTERY
DISEASE (SECONDARY PREVENTION):   AN APPROACH TO COST/BENEFIT
AND COST/EFFECTIVENESS ANALYSIS

L. Dardanoni,* P. Assennato,**
R. Oliveri,* and A. Raineri**

*Istituto di Igiene, Università di Palermo, Italy
**Cattedra di Fisiopatologia Cardiovascolare
Università di Palermo, Italy

Classification of different types of preventive measures for
ischemic heart disease (IHD) has been much debated. In the classical
use of terms "primary prevention" (prevention of occurrence) includes
removal of risk factors in otherwise healthy individuals, while the
attempts to change diet and life style in the very early age, in
order to avoid the acquisition of risk factors, has been called "pre-
primary prevention" or "early prevention."

"Secondary prevention" (prevention of progression) has been
defined as treatment of asymptomatic individuals having a substantial
and presumably progressive cardiac ischemia; this can be effected by
i) screening healthy individuals; ii) treatment of screened people.

Treatment of symptomatic cases (patients with angina or myocardial infarction) in order to prevent infarction or reinfarction
is sometimes defined as "tertiary prevention."

Since in existing literature different meanings are attributed
to these terms, it is unlikely that all will agree to the above
quoted definitions: however, for the scope of this lecture,
"secondary prevention" will indicate the application of a screening
test to asymptomatic individuals, followed by treatment of those in
which cardiac ischemia was demonstrated by the screening test or by
confirmatory procedures.

The possibility of a practical development of programmes of
secondary prevention of IHD has been increased both by the existence
of non invasive tests for the assessment of myocardial ischemia and
of effective treatment of such pathological conditions, which can
favourably modify the prognosis of ischemic patients[2,8,9,12] i.e.,

to reduce morbidity and mortality due to coronary artery disease and or increase life expectancy or quality of life.

The screening test could be in principle applied to the whole population, also in combination with the more usual tests for other risk factors of IHD; however, it is obvious that a selection of the population at risk or at high risk is advisable, and that the study of three main risk factors (cholesterol, blood pressure, cigarettes) might be simultaneously performed or might precede the screening for asymptomatic IHD; the latter could therefore be applied only to individuals having one or more risk factors. Furthermore, all individuals submitted to secondary prevention measures might as well undergo primary preventive treatment.

Therefore, since primary and secondary screening and treatment should usually be carried out at the same time, in the study of cost and benefit of the secondary prevention only marginal (incremental) inputs and outputs must be evaluated, i.e., the cost of resources used for the screening test itself and for treatment of silent ischemia; from the benefit side, the extra health gains attributable to the specific treatment, as distinguished from the gain connected with the compliance to the usual measures of risk factors removal or reduction (diet, life-style changes, blood pressure control.)

In order to carry out a cost/benefit (or cost/effectiveness) analysis of the programme, a certain amount of information is needed.

1) According to Cohn (1981)[2], detection of silent myocardial ischemia could be performed by Holter monitoring or by exercise test; the latter may be done at the maximal and submaximal level and the results may be measured by ECG, echocardiogram or radionuclide ventriculogram. For each testing procedure it is necessary to know sensitivity and specificity. In the case of a combination of procedures, sensitivity and specificity will be strongly influenced by the decision to screen out individuals when at least one test is positive or when both must be positive.

2) The prevalence of asymptomatic ischemia in the population, or respectively in low or high risk groups must be assessed. Acceptable data are not presently available, since most of the studies published up to now have been applied on biased subjects mostly partially symptomatic or recruited on a voluntary basis[2,4,10,13].

Also different aims have been adopted, going on a post mortem study of coronary narrowing on subjects deceased for non circulatory conditions[2,9] to coronarography. According to an estimate of Cohn[2] based on the data of Diamond and Foster (1979)[4] about 5% of middle aged men have sufficiently severe coronary narrowing; prevalence in women is estimated to be about 2.5%. It is obvious that this applies to North America; a very rough estimate of the prevalence in other

populations could be carried out if it is assumed that the ratio
myocardial ischemia/death for coronary disease is constant.

Prevalence must be evaluated for each risk level, in order to
calculate the predictive value of each screening procedure as applied
to high, medium or low risk groups. As for risk assessment, the well
known quantification adopted in several studies may be used[7,11].

Since treatment must be preceded by an accurate evaluation of
the extent of the coronary disease, a confirmatory test must be
applied to all positive subjects in the screening test. Therefore,
sensitivity and specificity of the second level test, (i.e., coronarography, contrast ventriculography, metabolic and hemodinamic studies,
coronary blood flow measurements) must also be assessed.

The most important date for the evaluation of the benefits (and/
or of the effectiveness) of secondary prevention concern the natural
history of asymptomatic ischemic patients, detected by the screening
and confirmatory test, in comparison to pair matched test negative
individuals it is particularly necessary to know mortality and inci-
dent morbidity due to coronary disease in both groups. As a general
health index, to be used for computation of health output attributable
to the secondary prevention, the well-life expectancy, or the well-
year (Fanshel and Bush (1972)[6], Epstein et al. (1981)[5]) could be
profitably calculated for negative test individuals from census data,
and for positive, from morbidity data and from various degrees of
disability connected to IHD.

The effectiveness of each type of treatment is only approximately
known. According to Cohn[2] cardiac surgery, beta blocking drugs and
exercise (or combination of such treatment) can be applied to silent
ischemia patients. Published data on the outcome of these different
interventions[8,12] mostly carried out on angina patients, cannot be
directly applied to estimate effectiveness in asymptomatic individuals; furthermore, on a long term evaluation, the average compliance
to acute (surgical) or cronic (drug, exercise) treatment should be
taken into account: it is known that lack of compliance is the main
cause of ineffectiveness of several prevention procedures highly
effective _per se_.

The same type of evaluation should be applied to the presumed
effectiveness of primary prevention measures, which will obviously
be carried out simultaneously on all subjects. Results of several
trials of the monofactorial or multifactorial type of intervention
are being published presently[7,11]. Some of them show that a substantial reduction of morbidity and mortality can be achieved through
health education methods; this can be considered a base-line beyond
which all the health gains of secondary prevention must be calculated
as a marginal (or incremental) gains.

Cost/effectiveness analysis may be relatively more simple than the cost/benefits one.

The cost of screening and of the second level confirmatory tests can be calculated either synthetically, through the market prices of each test, or analytically in terms of manpower, material, instruments. The latter procedure must be preferred, since prices of the tests cannot be directly obtained in areas in which medical care is more or less socialized. Time loss and discomfort of the screened population should also be taken into account. The cost of unwanted effects of the screening procedures must also be evaluated: although the proposed screening tests are non invasive, some incidents, severe enough to require hospitalization, have been described; more inconvenience can be expected from the second level testing.

Cost of each type of treatment (drugs, exercise, surgery) of screened individuals must be calculated, as well as cost of time loss, discomfort, treatment of complications, medical surveillance (particularly after surgery); calculation of cost in momentary units can be performed using the same procedures adopted for the screening; however, monetization of some items like discomfort or unwanted effects of treatments is an imperfectly solved problem. In any case an attempt to quantify such items must be performed.

The effectiveness of single treatment can be calculated classically as: 1) number of deaths due to IHD avoided; 2) number of IHD avoided; 3) reduction of average duration and degree of disability in IHD cases.

Since end-point data are strictly comparable, the cost/effectiveness of each combination of the screening procedure plus preventive treatment will be obtained for different aims, namely mortality, morbidity and disability.

In the cost/benefit analysis, while marginal costs are essentially those already mentioned, the outcome must be evaluated using the same monetary unit, since the analysis will compare the two alternatives of implementing the prevention programme or not. A suitable method for the measurement of health gain is the use of "weighting" procedure proposed by Fanshel and Bush (1970)[6] for the calculation of the Health Status Index, as applied to tuberculin testing, Bush et al., (1972), phenylketonuria, Bush et al., (1973)[1] and to abnormal serum tyroxine, Epstein et al., (1981)[5]. Briefly, according to Epstein et al., the outcomes are expressed in a single unit, the Well-Year, i.e., "the equivalent of years of life which are completely functional and asymptomatic." The health gain is therefore: $\Delta E = E_1 - E_0$ where E_1 is the symptom specific Weighted Life Expectancy with the diagnostic procedure/treatment, and E_0 is the symptom specific Weighted Life Expectancy without the diagnostic procedure/treatment." For the computation of Well-Year tables R_1 and R_2 of the Appendix by

Bush and Schneiderman attached to Epstein et al., paper must be consulted.

Since benefits are expected in the future, while costs are incurred both at the beginning and during the development of the program, discounting procedures must be adopted. The value of one Well-Year can be expressed, according to the above quoted Authors, in monetary units and compared with the cost of the preventive programme. It can be concluded that, although methods for cost/effectiveness and cost/benefit analyses can reasonably be defined, more epidemiological data must be collected for an acceptable application to secondary prevention of IHD

SUMMARY

Methods for the cost/effectiveness and cost/benefit analysis of a program of secondary prevention of ischemic heart disease are discussed.

Data for the evaluation of the cost of the program, as well as data on prevalence of asymptomatic coronary disease, related morbidity and mortality must be obtained through epidemiological investigations.

A list of requirements for the analysis is discussed.

REFERENCES

1. J.W. Bush, M.M. Chen, and D.L. Patrick, Cost effectiveness using a health status index: analysis of the New York State PKU screening program, in: "Health Status Indexes," R. Berg, ed., Chicago Hospital research and educational trust (1973).
2. P.E. Cohn, Asymptomatic coronary artery disease, Mod.Concepts Cardiovasc.Dis. 50:55 (1981).
3. S. Cretin, Cost/benefit analysis of treatment and prevention of myocardial infarction, Health Serv.Res. 12:174 (1977).
4. G.A. Diamond and J.S. Forrester, Analysis of probability as an aid in the clinical diagnosis of coronary artery disease, N.Engl.J.Med. 300:1350 (1979).
5. K.A. Epstein, L.J. Schneiderman, J.W. Bush, and A. Zettner, The "abnormal" screening serum tyroxine (T4): analysis of physician response, outcome, cost and health effectiveness, J.Chron.Dis. 84:175 (1981).
6. S. Fanshel and J.W. Bush, A health status index and its application to health services outcomes, Oper.Res. 18:1021 (1970).
7. Gruppo di Ricerca del Progetto Romano di prevenzione della cardiopatia coronarica. Il Progetto Romano di prevenzione delle cardiopatie coronariche: Risultati finali, Istituto di Sanità 12. (1982) (ISSN0391-1675).

8. K.E. Hammermeister, T.A. De Roven, and H.T. Dodge, Effect of
    coronary surgery on survival in asymptomatic and minimally
    symptomatic patients, Circulation 62 (Suppl I) 98 (1980).
9. J.R. Hickman Jr., G.S. Uhl, R.L. Cook, P.J. Engel, and A.
    Hopkirck, A natural history study of asymptomatic coronary
    disease, Am.J.Cardiol. 45:422 (1980).
10. C. Manca, L. DeiCas, D. Albertini, G. Baldi, and O. Visioli,
    Different prognostic value of exercise electrocardiogram in
    men and women, Cardiol. 65:312 (1978).
11. J.T. Salonen, P. Puska, T.E. Kottke, and J. Tuomilchto, Changes
    in smoking, serum cholesterol and blood pressure levels during
    a community-based cardiovascular prevention program. The North
    Karelia project, Am.J.Epid. 114:81 (1981).
12. L. Wilhelmsen, The use of beta-blockers in secondary prevention.
    Lecture, School of Cardiology, E. Majorana Center of Scientific
    Culture, Erice (1982) (This book, page 241).
13. L. Wilhelmsen, J. Bjure, B. Ekström-Jodal, M. Aurell, G. Grimby,
    K. Svärsudd, G. Tibblin, and H. Wedel, Nine years' follow-up
    of a maximal exercise test in a random population sample of
    middle aged men, Cardiology 68 (Suppl.2) 1 (1981).

PHYSICAL ACTIVITY IN THE PREVENTION

OF CORONARY HEART DISEASE*

Victor Froelicher

Cardiac Rehabilitation & Exercise Testing
University Hospital
School of Medicine
University of California, San Diego

INTRODUCTION

Epidemiologists have had difficulty documenting the health
benefits of exercise because of the problems with selection and
premorbid job transfers, quantifying activity levels, accurate diagnosis of coronary disease and the interaction with other risk predictors. In addition, the type of exercise that is beneficial and
its method of protection is uncertain. Since an exercise program
can lower catecholamine levels and increase fibrinolysis, the beneficial action could be mediated by hormonal and clotting factor
changes possibly not related to the hemodynamic changes produced by
an exercise program. In order to clarify this last problem, this
review will begin with a description of the body's acute and chronic
response to exercise. Following will be a summary of the studies
of Paffenbarger, Morris and colleagues which have demonstrated
physical inactivity to be a risk factor in three different apparently
well populations and have provided some quantification of protective
levels of activity. Fortunately, high levels of activity (i.e.,
marathoning) are not necessary and modest levels of exercise provide
a decreased probability of developing coronary disease. Recent
autopsy studies have identified the pathology related to the rare
event of sudden death during exercise in the apparently healthy
individual. Lastly, two recent exercise-intervention studies of
secondary prevention will be reviewed.

*Supported by: Specialized Center of Research on Ischemic Heart
Disease, NIH Research Grant HL 17682 awarded by the National Heart,
Lung and Blood Institute to John Ross, M.D.

THE ACUTE RESPONSE TO EXERCISE

The response to dynamic muscular exercise[1] consists of a complex
series of cardiovascular adjustments designed to: 1) see that active
muscles receive a blood supply appropriate to their metabolic needs;
2) dissipate the heat generated by active muscles; 3) maintain the
blood supply to the brain and the heart.

The regulation of the circulation during exercise involves the
four following adaptations: 1) Local - the resistance vessels dilate
in the active muscles owing to the products of muscle metabolism.
There is an immediate dilation of the arteries and arterioles in
active muscle because of the sudden increase in metabolites. These
products block the action of the sympathetic nerves on the muscle
vessels so that they do not constrict. This results in a decrease
in systemic vascular resistance proportional to the muscle mass
involved. 2) Nervous Adaptations - the sympathetic outflow to the
heart and systemic blood vessels is increased to maintain arterial
blood pressure; the vagal outflow to the heart decreases. This
causes tachycardia, increased cardiac contractility, and constriction
of the resistance vessels in the kidneys and the gut. The resistance
vessels also constrict in non-working muscles. The increased sympathetic outflow is due in part to a central command from the
cerebral cortex and to activation of chemoreceptors in contracting
skeletal muscles. The arterial and cardiopulmonary baroreceptors
prevent marked fluctuations in systemic pressure from normal values.
As exercise continues and body temperature rises, the temperature
sensitive cells in the hypothalmus are activated. They inhibit the
sympathetic outflow to the skin vessels and stimulate the cholinergic
fibers to the sweat glands. This results in dilation of the skin
vessels and increased sweating. 3) Humoral Adaptations - if exercise
is severe, the cholinergic fibers to the adrenal medulla are activated and epinephrine is released into the blood stream. This
further increases the heart rate and myocardial contractility, and
tightens the constriction of the veins and the renal arterial system.
4) Mechanical Adaptations - during exercise, the active muscles in
the legs return blood from the legs to the central circulation using
the one-way valves in the leg veins. As cardiac output increases,
there is an increase in systemic arterial pressure. The increase
in pulmonary blood flow causes a moderate increase in mean pulmonary
artery pressure. The relationship of pressure, flow and resistance
is defined in OHM's law: this physical law states that resistance
creases in the tissues that do not function in the performance of
the ongoing exercise and decreases in active muscle. The total
results is a decrease in overall systemic resistance. This is explained by the fact that while pressure only increases mildly, flow
can increase by as much as five times during dynamic exercise.
Since the denominator (flow) increases much more than the numerator
(pressure) in the formula of resistance, the result is a decrease
in systemic resistance. Another mechanical adaptation occurs when

the increasing venous return dilates the left ventricle and cardiac
function is enhanced via the Frank-Starling mechanism.

There is a highly predictable relationship between the total
body oxygen consumption and both the cardiovascular and respiratory
responses to exercise. To explain this relationship, six major
hypotheses have been advanced.[1] First, the arterial baroreflex
hypothesis - this is based on the idea that vasodilation of active
muscle causes a fall in blood pressure which in turn would trigger
a baroreflex and raise heart rate and cardiac output. However, a
fall in blood pressure does not initiate the exercise response.
Second, the central nervous system excitation hypothesis - the outflow of motor impulses could interact with the centers that regulate
the cardiovascular responses to exercise. The major problem with
this hypothesis is that there is no feedback mechanism to the central
nervous system to maintain the delicate relationship between these
responses and exercising muscle. The third and fourth hypotheses
are based on chemoreflexes in the arterial or central venous systems.
However, there is little data to support the idea that changes in
PO_2, CO_2, or PH are mediators. Fifth, the skeletal muscle mechano-
receptors hypothesis - these receptors cannot be involved in the
exercise reflex since there is no cardiovascular or respiratory
response to muscle vibration which is a potent stimulus to mechano-
receptors and selective blockade of large mechanoreceptor afferents
does not block the exercise response. The sixth and most logical
hypothesis is based on muscle chemoreceptors. The most current
evidence suggests that some sensor within skeletal muscle, detects
small changes in the local chemical environment and serves to monitor
the adequacy of muscle perfusion. These receptors then transmit
their signals over neural pathways to the central nervous system.

THE RESPONSE TO CHRONIC EXERCISE

Chronic exercise or an exercise program has also been called
"Training" or "Physical Conditioning". What is meant is that an
individual maintains a regular habit of exercise at levels greater
than he or she usually performed. Exercise can be designed for
increasing muscular strength or dynamic (aerobic) performance. The
type of exercise that results in an increase in muscular strength
is isometric exercise or exerting muscular tension with little
movement against resistance. Though this results in an increase in
muscular mass along with strength, such exercises have little effect
on the cardiovascular system. The heart works against a pressure
load without much of an increase in cardiac output. Dynamic exercise,
also called isotonic or aerobic, involves the rapid movement of
large muscle masses that results in the need for the body to respond
with increasing ventilation in order to increase oxygen consumption.
Such exercise is also called "aerobic" since it must be performed
without accumulating an anaerobic debt. The heart must increase its

output and performs flow work rather than pressure work. This is the type of exercise that results in the cardiovascular changes that will be described.

The features of an aerobic exercise program that must be considered include: the mode, the duration, the intensity, and the frequency of exercise.[2] In general, the mode of exercise must involve movement of large masses of muscle. Such exercise includes bicycling, walking, running, and swimming. The exercise should be carried out in at least three sessions a week and be inter-spaced throughout the week. Duration should be 30 minutes to one hour. Intensity should be 50% or greater of the maximal oxygen consumption and involve at least 300 k/cal of energy expenditure. The percentage of maximal oxygen consumption being performed can be approximated from the heart rate and perceived exertion.[3]

The results of such an aerobic exercise program can be grouped as: 1) hemodynamic; 2) morphologic; and, 3) metabolic.[4] The hemodynamic consequences of an exercise program include a decrease in resting heart rate, a decrease in the heart rate and systolic blood pressure at any matched submaximal workload, an increase in work capacity and maximal oxygen consumption, and a faster recovery from an exercise bout. It is argued whether these changes are due to peripheral or to central adaptations but probably they are due to both. Peripheral adaptations are more important in older individuals and in patients with heart or lung disease, while central changes are probably more of a factor in younger individuals. Central hemodynamic changes that have been observed in some instances include enhanced cardiac function and cardiac output.

The morphologic changes that occur with an exercise program are clearly age-related. These changes occur most definitely in younger individuals and may not occur in older individuals. The exact age at which the response to chronic exercise is altered is uncertain, but it would seem to be in the early 30's. Morphologic changes include an increase in myocardial mass and left ventricular end-diastolic volume. Paralleling these changes are increases in coronary artery size and in the myocardial capillary to fiber ratio. These changes are clearly beneficial, making it possible for the heart to function better under any stress and also to be better perfused during any stress. In older individuals, there might even be a decrease in myocardial mass that still results in an improvement in capillary to muscle fiber ratio, but no change in coronary artery size. No studies have shown a decrease in atherosclerotic plagues once they are present with an exercise program. However, a recent monkey study has shown that exercise may offset the impact of an atherogenetic diet to some degree.

Kramsch and colleagues randomly allocated 27 young adult male monkeys into three groups.[5] Two groups were studied for 36 months

and one group for 42 months. One of the groups studied for 36
months received a vegetarian diet for the entire study while the
other group received it for 12 months then was fed an isocaloric
atherogenetic diet for 24 months. Both were limited in activity to
that permitted by a single cage and were therefore designated as
sedentary. The third group ate the vegetarian diet for 18 months
and were exercised on a treadmill regularly. Then they were given
the atherogenetic diet for 24 more months while the exercise program
was continued. Two of the monkeys on the atherogenetic diet, one
sedentary and one exercising, had no elevation of serum cholesterol
and so they were eliminated from the study. Over three years, the
animals were observed for objective evidence to support the protective value of exercise. Total serum cholesterol was the same,
but HDL cholesterol was higher in the exercise group. Ischemic
electrocardiographic changes, significant angiographic coronary
artery narrowing, and sudden death were observed only in the
sedentary monkeys fed the atherogenetic diet. In addition, post-
mortem examination revealed marked coronary atherosclerosis and
stenosis. Exercise was associated with substantially reduced
overall atherogenetic involvement, lesion size and collagen accumulation. Their results showed that exercise in young adult monkeys
increased heart size, left ventricular mass, and the diameter of
coronary arteries. Also, the subsequent experimental atherosclerosis
induced by the atherogenetic diet given for two years was substantially reduced. Thus, an exercise program before initiating an
atherogenetic diet delayed the manifestations of coronary heart
disease. Of importance is at what point comparable to human life-
span were these studies initiated and what percentage of that span
was represented by the three years of observation.

Animal studies have provided substantial evidence of the cardiovascular benefits of regular physical activity[6]. Improved coronary
circulation has been demonstrated in exercise trained animals by
increased coronary artery size, capillary density, reduced myocardial
infarction size, and maintenance of coronary flow in response to
hypoxia. However, whether changes in myocardial collateral perfusion
occurs remains controversial, but exercise probably improves this
when ischemia is present. Studies utilizing various animal models
have reported improvement in cardiac function secondary to exercise
training. Improved intrinsic contractility, faster relaxation,
enzymatic alterations, calcium availability and enhanced autonomic
and hormonal control of function have all been implicated. Perhaps
the beneficial effects of exercise would be more apparent in humans
if we were as compliant to an exercise program as animals.

Perhaps behavioral scientists should help people become
"addicted" to exercise.[7] This has been defined as addiction, of
a psychological and/or physiological nature, to a regular habit of
exercise, characterized by withdrawal symptoms if not performed
every one or two days. These withdrawal symptoms can include

anxiety, restlessness, guilt, irritability, tension, bloatedness, muscle twitching and discomfort. Such an addiction probably does not occur until after two years or more of running.

Echocardiography has been utilized to evaluate cardiac adaptations to exercise training in longitudinal studies.[4] Reported cardiac changes secondary to endurance training in young subjects have included increased ventricular mass, wall thickness, and volume. However, echocardiographic measurements of other groups of young subjects and older subjects failed to document these changes inspite of improvement in maximal oxygen consumption. These echocardiographic studies suggest that increase in left ventricular mass may not occur in younger subjects unless higher levels of training are used and may never occur in older subjects, particularly those with coronary disease. The studies of the effects of an exercise program on the abnormal exercise electrocardiogram have also been controversial.

The metabolic alterations secondary to an aerobic exercise program are summarized below. Total serum cholesterol level is not effected, but HDL levels are increased. These are the lipoproteins that remove cholesterol from the body. Serum triglyceride and fasting glucose levels are decreased. In addition, it appears that there are favorable alterations in insulin and glucagon responses. Diabetics need less insulin if they maintain a regular exercise program. Also, after a training program, catecholamine levels are lower in response to any stress. The fibrinolytic system seems to be enhanced and since coronary thrombosis is no longer a misnomer, this would seem to be beneficial in preventing myocardial infarction.

Though frequently comments are made regarding exercise enhanced psychological well-being and the "runners high", few scientific studies have been performed in this area. However, it would seem that exercise does have a tranquilizing effect and increases pain tolerance which may be beneficial in some individuals.

EPIDEMIOLOGIC STUDIES

Epidemiologic data must be evaluated using established guidelines to ascertain whether the relationship of a factor and a disease is causal or casual. Three recent studies have satisfied these guidelines.

Epstein, Morris and colleagues studied the relationship of vigorous exercise during leisure time to the resting electrocardiogram.[8] On a randomly selected Monday morning approximately 17,000 middle-aged men recorded their leisure activities of the previous weekend. Their work was sedentary and other pertinent risk factors

were also assessed. Vigorous exercise during leisure time was
reported by 25%, and these active men had significantly fewer
electrocardiographic abnormalities. The only demonstrated relationship found was between increased electrocardiographic abnormalities
and elevated blood pressure. No relationship was found between
electrocardiographic abnormalities and serum cholesterol levels or
cigarette smoking. However, an 8½ year follow-up demonstrated half
as much coronary heart disease in those who reported vigorous
leisure activity.[9]

Paffenbarger and co-workers analyzed a 22 year follow-up of
longshoremen from 1951 to 1972, a total of 59,401 person-years, for
a relationship between coronary heart disease and energy expenditure
on the job.[10] Little difference was found in other risk factors
when the high and low activity groups were compared. Low energy
output, smoking more than one pack of cigarettes a day, and systolic
blood pressure equal to or greater than the mean all identified men
at increased risk for a fatal heart attack. Hypothetically, if all
the men had transferred to high activity jobs, the number of heart
attacks among them would have been reduced by one-half.

In a second study Paffenbarger and co-workers reported a follow-
up of 36,000 Harvard University alumni who had entered college
between 1916 and 1950.[11] Records of their physical activity were
gathered from their student days and during middle-age. A six to
ten year follow-up from 1962 to 1972 included 117,680 person-years.
After the subjects completed the first questionnaire, apparently
healthy men were classified by a physical activity index that considered stairs climbed, blocks walked, and sports played. Division
of this index at 2,000 kcal/week (roughly equivalent to running two
miles a day) produced a 60% to 40% division of person-years of
observation into low and high energy categories, respectively.

During the six to ten year follow-up, 572 men had their first
myocardial infarction. Men with a low physical activity index were
at 64% higher risk than were classmates with a high activity index.
Former varsity athletes retained a lower risk only if they maintained a high physical activity index (greater than 2,000 kcal/week).
There was a 50% increase in risk with the presence of any one risk
factor, and the presence of two tripled the risk. Maintenance of
a high physical activity index might have reduced the heart attack
rate by 26%.

Kannel and Sorlie analyzed the Framingham data for the effects
of physical activity on overall mortality and cardiovascular disease
mortality.[12] The effect on mortality of being sedentary was modest
compared with the effect of other risk factors, but it persisted
when these other factors were taken into account. A low correlation
was noted between physical activity level and the major risk factors.

DEATH AND OTHER COMPLICATIONS OF EXERCISE

Recent reports have disproven the hypothesis that marathon
running provides absolute protection against dying from coronary
atherosclerosis. Waller and Roberts have reported the autopsies of
five conditioned runners aged 40 years and over, all with severe
coronary atherosclerosis.[13] These researchers concluded that coronary heart disease is a major killer of conditioned runners aged
40 years and over who die while running. Thompson and colleagues
investigated the circumstances of death by considering the medical
and activity histories of 18 individuals who died during or immediately after jogging and found coronary heart disease to be the most
common cause.[14]

What is the expected level of cardiovascular deaths among
runners while running on the basis of chance alone? This is an
important question because it is frequently assumed that exercise
is the cause when a person dies of cardiovascular causes during
recreational running. Koplan used data from the National Center
of Health Statistics and found that approximately 100 cardiovascular
deaths per year in runners in the United States can be predicted on
a purely temporal basis.[15] This is certainly higher than the number
of deaths reported.

Morales and colleagues have reported three healthy individuals
who died suddenly during strenuous exercise, and were found to have
a triad of pathological findings.[16] There were two males aged 34
and 54 and one female 17 years of age. The pathological triad was
muscle bridging of the left anterior descending coronary artery,
poor circulation to the posterior surface of the heart, and septal
fibrosis. The angiographic finding of a coronary artery that passes
underneath a band of myocardium is not that unusual and it has been
debated whether or not it has functional significance. Some studies
of coronary blood flow have suggested that the constriction of a
coronary artery by this myocardial band during systole results in
decreased flow, however, most of coronary flow takes place during
diastole. In regard to the second finding, there is great variability
in the coronary artery distribution on the posterior surface of the
heart around the crux and the posterior margin of the septum. In
the most common situation, the right coronary artery branches into
a posterior descending artery which passes down the septum giving
off septal perforators. Often though, there are normal variations
where the left circumflex provides this branch or there are only
small arteries in the area. Lastly, septal fibrosis could be due
to chronic ischemia. In summary, these anatomic findings are not
specific for inadequate myocardial blood flow and they could be
purely coincidential in these three individuals.

Noakes and colleagues have presented four marathon runners with
autopsy-proven coronary atherosclerosis.[17] The first individual

was a 44 year old white male who, after 14 months of training, had
completed seven marathons in under four hours. He suddenly dropped
dead half way through a marathon. At autopsy he was found to have
an old anteroseptal myocardial infarction and 90% lesions of his
left anterior descending and circumflex coronary arteries. The
second was a 41 year old male, who after two years of running, had
a symptomatic myocardial infarction. After release from the hospital,
he returned to training and ran in five marathons. He was hospitalized with unstable angina and coronary angiography was performed.
He was found to have severe triple vessel coronary artery disease;
while waiting for surgery, he died suddenly. The last two cases
were 36 and 27 year old athletes who had completed multiple marathons
and were killed accidentally. Both had left anterior descending
coronary artery lesions at autopsy; the younger a 50% and the older
a 90% lesion.

Maron, Roberts and colleagues from the National Institute of
Health have reported more than 30 athletes ranging in age from 13
to 30 who died suddenly.[18] Over half of them died in proximity to
the athletic field. The most common cause of death was hypertrophic
cardiomyopathy. Several however, died of coronary atherosclerosis,
one after running a pass pattern in a professional football game.
In this individual they hypothesized that a blow to the chest while
being tackled caused a hemorrhage into a plaque in the left anterior
descending coronary artery. Several others had congenital anomalies
of the coronary arteries. All of these events are extremely unusual
and it would be difficult to screen for them. It is known that
athletes frequently have abnormal electrocardiograms and even echocardiographic hypertrophy. In addition, they have a higher prevalence
of false-positive exercise tests. However, screening for lipid
abnormalities would be a wise public health measure regardless of
a lack of specificity and it would be advisable to get an echocardiogram on an athlete with symptoms or signs of a hypertrophic
cardiomyopathy.

Short of death, there are numerous other risks for amateur
and professional athletes. Heat stroke can be avoided by taking
precautions for high humidity, high temperature environments including drinking adequate replacement amounts of dilute electrolyte
solutions. Hopefully, there are no longer coaches who restrict
fluids in order to limit sweating. Interestingly, a recent report
has shown that runners can have heat stroke and still be actively
sweating, while once it was taught that heat stroke was always
preceded by a cessation of sweating.[19] Hematuria after a run can
be due to bladder trauma, but proteinuria appears to be due to
renal ischemia.[20] Diarrhea and other gastro-intestinal complaints
are fairly common in runners during and after events.[21] Numerous
episodes of anaphylaxis thought to be exercise-induced have been
reported.[22] The importance of diagnosing this by the findings of
bronchospasm and urticaria, is that treatment with epinephrine and

anti-histamines can be lifesaving. This usually occurs in the
individuals that previously had an anaphylactic reaction to shell-
fish. The many different types of orthopedic injuries that can
occur are too numerous to list here.

RECENT STUDIES OF EXERCISE IN THE SECONDARY
PREVENTION OF CORONARY HEART DISEASE

Ehsani and colleagues have reported results of one year of
intense aerobic exercise training in a highly selected group of 10
patients with coronary heart disease.[23] The patients, ranging in
age from 44 to 63, were the first in their program to complete 12
months in a high level exercise program. Nine had sustained a
single myocardial infarction while the other had severe three vessel
coronary artery disease. All 10 had asymptomatic, exercise-induced
ST segment depression. Eight similar men were considered as
controls. After three months of exercise training at a level of
50% to 70% of their maximal oxygen consumption, they were allowed
to increase their level to 70% to 80%, with two to three intervals
at 80% to 90% interspersed throughout the exercise session. Patients
exercised three times per week during the first three months, and
four to five times per week for the next nine months. The duration
was at first 30 minutes and later was increased to 60 minutes. No
complications were reported. Since none of these men developed
symptoms during exercise testing a true maximal oxygen consumption
could be measured. The maximal amount of reported ST segment
depression was .30 mv, but most had .20 mv of depression which was
less at repeat testing one year later inspite of a higher double
product, greater treadmill workload, and a 38% greater maximal oxygen
consumption. In addition, 0.1 mv of ST segment depression occurred
at a higher double product after the year of training. A weight
loss from a mean of 79 kg to 74 kg occurred. The sum of SV1 and
RV5 increased by 15%. Both left ventricular end diastolic dimension
and posterior wall thickness were significantly increased after
training. This resulted in an increase in left ventricular mass
from 93 to 135 grams per meter squared body surface area. One
wonders if ST segment depression in Z was included since the specific
lead with ST depression was not given. In Z, ST depression is really
ST elevation anteriorly and is probably not due to ischemia. Also,
how many of these asymptomatic men with coronary disease actually
had false positive ST abnormalities - that is, depression not due
to ischemia?

Though these results are exciting, the relevance to clinical
practice is uncertain. These 10 men are a highly selected group,
all with asymptomatic ST depression and able to exercise at levels
often difficult for younger men. If applied to most patients with
ischemia heart disease, this intensity certainly would lead to a

high incidence of orthopedic and cardiac complications. Also,
similar studies in cardiac patients have failed to confirm their
results.[24,25]

The National Exercise and Heart Disease Project included 651
post-myocardial infarction men enrolled in five U.S. centers.[26] It
was a randomized three year clinical trial of the effects of a
prescribed supervised exercise program starting two to 36 months
post-myocardial infarction (80% were more than eight months post).
Three hundred and twenty-three randomly selected patients underwent
exercise three times a week designed to increase their heart rate
to 85% of the individual maximal heart rate during treadmill testing,
and 328 patients served as controls. This study was carefully performed by experts who took two years to design the protocol. An
initial low level exercise session in both groups to exclude the
"faint of heart" who would not comply with an exercise program was
surprisingly effective in improving performance.

The three year mortality rate was 7.3% (24 deaths) in the
control group versus 4.6% (15 deaths) in the exercise group. Deaths
from all cardiovascular causes (acute myocardial infarction, sudden
death, arrhythmias, congestive heart failure, cardiogenic shock,
and stroke) were 6.1% per three years (20 deaths) in the control
group versus 4.3% (14 deaths) in the exercise group. Neither difference was statistically significant. However, when deaths due to
acute myocardial infarction were considered as a separate category,
the exercise group had a significantly lower rate: one acute fatal
myocardial infarction per three years (0.3%) in the exercise versus
eight fatal myocardial infarctions (2.4%) in the control group
(p < .05). The rate of all recurrent myocardial infarctions per
three years (fatal plus nonfatal) did not significantly differ
between groups (23 cases in the control versus 17 cases in the
exercise group, 7.0% versus 5.3%). The number of rehospitalizations
for reasons other than myocardial infarction were identical in the
two groups (27.4% versus 28.5% per three years). The need for
coronary artery surgery was also equal in both groups; 16 controls
and 17 patients in the exercise group underwent surgery during the
three year period.

Insufficient enrollees due to financial limitations and dropouts
prevented a definitive conclusion, but this study demonstrated the
feasibility of resolving the question of the value of exercise
training post-myocardial infarction. It is most unfortunate that
it was discontinued especially since the results are so encouraging.
Only 1,400 patients would be required to demonstrate a statistically
significant reduction in mortality rate in the exercise group if
the reported trend persisted. The patients in the exercise group
who suffered recurrent myocardial infarctions had a lower mortality
rate, suggesting that an exercise program increases an individual's
ability to survive a myocardial infarction.

PERFEXT

Exercise training has multiple potential ways that it could benefit the cardiac patient. There could be benefits due to alterations in cardiac physiology, but there are no definitive studies in cardiac patients that demonstrate improvement in cardiac function and perfusion due to exercise training. The new radionuclide techniques that can be performed during exercise give us the opportunity to investigate this question. Thallium treadmill testing[27] and supine bike radionuclide ventriculography[28] can be used to evaluate myocardial perfusion and function. If changes in the heart do occur with exercise training they are probably subtle and will be apparent only during the stress of exercise and not at rest. Therefore, nuclear cardiology techniques should be able to assess such changes particularly since they can be performed during exercise. Thallium-201 and Technetium-99m are radionuclides that give off radioactive energy that can activate the crystal of a standard scintillation camera. Thallium behaves like potassium and is taken up by viable, perfused myocardium. Areas of scar or ischemia produce "cold spots" in the scintillation image. Technetium is bound to red cells and by using a computer to collect picture frames time-related to the electrocardiogram, left ventriculographic images can be visualized. Several groups have initiated pilot studies of these techniques to evaluate cardiac rehabilitation. At the University of California in San Diego, we have initiated a randomized trial utilizing these techniques before and after one year. The trial is called PERFEXT, for perfusion-performance-exercise trial, and will be completed when 200 patients have been randomized and studied.

CONCLUSIONS

Although many clinicians exercise themselves and prescribe exercise for health reasons, the association between physical inactivity and the underlying atherosclerotic process is modest compared with that of other factors such as serum cholesterol levels, cigarette smoking, and hypertension. An inversely proportional association between the level of activity and degree of atherosclerosis has not been demonstrated. Physical inactivity does not necessarily precede the atherosclerotic process and the consistency of the exercise hypothesis has not been documented in autopsy studies. However, recent studies of primary and secondary prevention support a life-style that features regular physical activity. Regular exercise probably decreases the risk for coronary heart disease and helps to decrease other risk factors. Also, the risk of catastrophe due to moderate exercise is small in the apparently healthy individual. Excessive levels of exercise such as marathoning are neither necessary nor absolutely protective.

Regular moderate exercise makes good sense for many reasons. It can improve the quality of life by lessening fatigue and by increasing physical performance. The recommendation of a moderate exercise habit encourages people to pay attention to their health and make the changes necessary to lessen coronary risk factors. The most significant advances in public health have been in the prevention, not in the treatment of disease. The current interest in physical fitness may prove more effective than the medical profession in making the public take responsibiltiy for health maintenance.

REFERENCES

1. L. B. Rowell, What signals govern the cardiovascular responses to exercise? Med. Sci. Sports Exer., 12:307 (1980).
2. M. L. Pollock, The recommended quantity and quality of exercise for developing and maintaining fitness in healthy adults: Position statement of the American College of Sports Medicine, J. Card. Rehab., 1(5):375 (1981).
3. M. C. Gutmann, R. W. Squires, M. L. Pollock, C. Foster and J. Anhulm, Perceived exertion-heart rate relationship during exercise testing and training in cardiac patients, J. Card. Rehab., 1(1):52 (1981).
4. V. F. Froelicher, "Exercise - Clinical Concepts", Le Jacq Publishing Inc., New York (1982).
5. D. M. Kramsch, A. J. Aspen, B. M. Abramowitz, T. Kreimendahl, and W. B. Hood, Reduction of coronary atherosclerosis by moderate conditioning exercise in monkeys on an atherogenic diet, N. Engl. J. Med., 305:1483 (1981).
6. V. F. Froelicher and P. Brown, Exercise and coronary heart disease, J. Card. Rehab., 4:277 (1981).
7. M. L. Sachs and D. Pargman, Running addiction: A depth interview examination, J. Sports Behavior, 23:143 (1979).
8. L. Epstein, G. J. Miller, F. W. Stitt and J. N. Morris, Vigorous exercise in leisure time, coronary risk-factors, and resting electrocardiogram in middle-aged male civil servants, Br. Heart J., 38:4403 (1976).
9. J. N. Morris, R. Pollard, M. G. Everitt, S. P. W. Chave and A. M. Semmence, Vigorous exercise in leisure time: Protection against coronary heart disease, Lancet., 2:1207 (1980).
10. R. S. Paffenbarger, R. J. Brnad, R. I. Sholtz and D. L. Jung, Energy expenditure, cigarette smoking, and blood pressure level as related to death from specific diseases, Am. J. Epidemiol., 108:12 (1978).
11. R. S. Paffenberger, A. L. Wing and R. T. Hyde, Physical activity as an index of heart attack risk in college alumni, Am. J. Epidemiol., 108:161 (1978).
12. W. B. Kannel and P. Sorlie, Some health benefits of physical activity: The Framingham study, Arch. Intern. Med., 139:857 (1979).

13. B. F. Waller and W. C. Roberts, Sudden death while running in conditioned runners aged 40 years or over, Am. J. Cardiol., 45:1291 (1980).

14. P. D. Wood, Death during jogging or running, JAMA., 242:2578 (1979).

15. J. P. Koplan, Cardiovascular deaths while running, JAMA., 242:2578 (1979).

16. A. R. Morales, R. Romanelli and R. J. Boucek, The mural left anterior descending coronary artery, strenuous exercise and sudden death, Circulation, 62:230 (1980).

17. T. D. Noakes, L. H. Opie, A. G. Rose, P. H. T. Kleynhans, N. J. Schepers and R. Dowseswell, Autopsy-proved coronary atherosclerosis in marathon runners, N. Engl. J. Med., 301:86 (1979).

18. B. J. Maron, W. C. Roberts, H. A. McAllister, D. R. Rosing and S. E. Epstein, Sudden death in young athletes, Circulation, 62:218 (1980).

19. P. G. Hanson and S. W. Zimmerman, Exertional heatstroke in novice runners, JAMA., 242:154 (1979).

20. N. J. Blacklock, Bladder trauma in the long-distance runner: "10,000 metres haematuria", Br. J. Urol., 49:129 (1977).

21. R. N. Fogoros, Runner's trots: Gastrointestinal disturbances in runners, JAMA., 243:1743 (1980).

22. A. P. Kaplan, S. F. Natbony, A. P. Tawil, L. Fruchter and M. Foster, Exercise-induced anaphylaxis as a manifestation of cholinergic urticaria, J. Allergy Clin. Immunol., 68:319 (1981).

23. A. A. Ehsani, G. W. Heath, J. M. Hagberg, B. E. Sobel and J. O. Holloszy, Effects of 12 months of intense exercise training on ischemic ST-segment depression in patients with coronary artery disease, Circulation, 64:1116 (1981).

24. K. Watanabe, B. Bhargava and V. Froelicher, Computerized approach to evaluating rest and exercise-induced ECG/VCG changes after cardiac rehabilitation, Clin. Card., 5:27 (1982).

25. R. Ditchey, J. Watkins, M. McKirnan and V. Froelicher, Effects of exercise training on left ventricular mass in patients with coronary heart disease, Am. Heart J., 101:701 (1981).

26. L. W. Shaw, Effects of a prescribed supervised exercise program on mortality and cardiovascular morbidity in patients after a myocardial infarct., Am. J. Cardiol., 48:39 (1981).

27. D. Jensen, J. E. Atwood, V. Froelicher, D. McKirnan, W. Ashburn and J. Ross, Improvement in ventricular function during exercise following cardiac rehabilitation, Am. J. Cardiol., 46:770 (1980).

28. V. F. Froelicher, D. G. Jensen, J. E. Atwood, M. D. McKirnan, K. Gerber, R. Slutsky, A. Battler, W. Ashburn and J. Ross, Cardiac rehabilitation: Evidence for improvement in myocardial perfusion and function, Arch. Phys. Med. Rehab., 61:517 (1980).

PREVENTION OF SUDDEN DEATH: SELECTION OF PATIENTS AT RISK

Jeffrey S. Borer[*], David Miller,
Paul Phillips, Jeffrey W. Moses,
Harvey Goldberg[**], and Jeffrey Fisher

Cardiology Division
New York Hospital - Cornell Medical Center
New York, N.Y., U.S.A.

Prevention of sudden death requires identification of patients at risk, elucidation of the pathophysiology of the lethal event, and definition of appropriate therapy. As yet, achievement of these three goals is incomplete. Patients generally do not undergo currently available testing procedures (which are at best not totally predictive) until potential risk is heralded by some clinical event. Thus, patients with clinically silent or asymptomatic electrical instability or with coronary artery/ischemic heart disease without angina pectoris may not enter the health care system before an episode of sudden death. Moreover, even if descriptors of risk are noted prospectively, the pathophysiology of sudden lethal events is incompletely understood. The nature and clinical importance of subcellular biochemical abnormalities leading to cardiac electrical instability largely are unknown, and the relationship between chronic, irreversible electrical instability and active ischemia in patients with coronary artery disease remains unclear. Finally, even when risk and likely mechanism of death both are identified in an individual patient, current pharmacological and surgical therapies offer incomplete protection against the lethal event.

Nonetheless, during the past decade, major advances have occurred in our knowledge of the genesis of sudden death, of the nature of

---

*Dr. Borer is an Established Investigator of the American Heart Association.
**Dr. Goldberg is the Isadore Rosenfeld Heart Foundation Fellow of the Cornell University Medical College.

its victims, and of effective means of prophylaxis. Perhaps the greatest insights have been gained in the identification of the patient at risk.

I. ISCHEMIC HEART DISEASE

Numerically, by far the largest group at risk of sudden death comprises patients with ischemic heart disease. While primary electrical instability leading to sudden arrhythmic death is known to occur even in the absence of coronary artery disease, the association of sudden death with acute ischemic events remains a major public health concern.

A. SURVIVORS OF RECENT MYOCARDIAL INFARCTION

Though the presence of clinical evidence of ischemic heart disease defines a group at relatively high risk for sudden death as compared with the general population, subgroups at particularly high risk can be readily distinguished. For example, the patient who has survived a recent acute myocardial infarction manifests an average 15% mortality risk during the ensuing 6 to 12 months; the great majority of such deaths are sudden.

Several methods have been employed in recent years to select those patients at highest risk following infarction, for whom intensive pharmacologic and, perhaps, surgical therapy might be of benefit. Of these, the most promising approaches appear to involve exercise testing, radionuclide imaging, and, perhaps, formal invasive electrophysiologic testing.

1. Exercise Testing

While criteria based on symptom development or exercise duration during standard exercise testing have been found to be relatively poor predictors of short term mortality, the induction of ventricular arrhythmias during formal exercise testing within 3 weeks after infarction appears to have important prognostic value[1]. Thus, Theroux et al. found risk of sudden death to be 2.5-fold greater in patients with treadmill exercise-induced ventricular premature contractions than in patients who did not manifest this abnormality[2], and Ericsson et al. found that 3 month post infarction mortality was 4.3 times greater in the subgroup of patients manifesting exercise-induced ventricular extrasystoles than in patients devoid of such arrhythmia[3]. In contrast, however, other studies suggest either that exercise-induced arrhythmias provide no prognostic information[4] of that arrhythmia-induction during exercise at most modifies prognosis based on more powerful descriptions of risk[5].

Much recent information indicates that the ECG ST segment response during exercise is particularly useful in identifying patients at high risk of sudden death early after infarction. Thus, in a small group, Sami et al. found that all patients who died suddenly within 2 years of infarction manifested $\geqslant$.2mV ST segment depression during exercise early after infarction, while this abnormality was present in only a third of patients without coronary events during followup[6]. Among the 210 patients studied by Theroux et al. 91% of sudden deaths during the year after infarction occurred in the subgroup of 64 patients manifesting .1mV ST segment depression during exercise within 3 weeks after infarction[2]. Finally, analysis of the correlation of ST segment response, arterial pressure response and symptomatic response during exercise may be more predictive of sudden death risk than is analysis of any single variable[7].

In the peri-infarction setting, myocardial perfusion scintigraphy with thallium 201 also appears to be of value in selecting patients at high risk of sudden death during the ensuing 6-12 months. In 42 selected patients studied within 15 hours of the onset of symptoms of infarction, Silverman et al. found that 6-month post infarction mortality was 62% among 13 patients with "large" perfusion defects, while 6-month mortality was only 7% among 29 patients with smaller defects[8]. Similarly, in a recent brief report, Becker et al. employing quantitative methods to assess perfusion scans in 42 selected patients studied within 15 hours of infarction, found that large perfusion defects (seen in 13 patients) identified patients who manifested a 62% 6-month mortality, while small defects (seen in 30 patients) were associated with a 5% 6-month mortality[9]. Thus, myocardial perfusion scintigraphy, even when performed at rest alone, appears to be a potentially valuable predictor of sudden death risk in the peri-infarction setting. As yet, the additional prognostic information available during scintigraphy with exercise has not been assessed.

The prognostic value of assessment of left ventricular function at rest by radionuclide cineangigraphy also has been demonstrated in survivors of recent infarctions. Battler et al. found that determination of ejection fraction 1 to 4 days following infarction permitted detection of patients at relatively high risk for death or debility during the ensuing year, with death or pulmonary congestion developing within 1 year after infarction in 46 of 66 patients (70%) with ejection fraction <52%. Such morbid or lethal developments occurred in only 7 of 30 patients (23%) with ejection fraction $\geqslant$52% early after infarction[10]. Similarly, in studies performed 2 weeks after infarction, Shulze et al, found that death occurred within 7 months after infarction in 8 of 45 patients (18%) with ejection fraction <40% two weeks after infarction, while no deaths occurred among the 36 patients with ejection fraction $\geqslant$40%[11]. Twenty-four hour ambulatory electrocardiograms were obtained at the time of radionuclide angiography, but while the presence of complex ventricu-

lar arrhythmias indicated an increased mortality risk for those
patients with ejection fraction <40%, the presence of complex ventricular arrhythmias provided no additional prognostic information
for patients with ejection fraction ≥40%. Our own earlier results
corroborated those of Shulze et al. in that among 45 patients studied
2 to 3 weeks following infarction, death occurred during the ensuing
year only in patients with ejection fraction 35% at rest (4 deaths
among 13 patients (31%))[12]. However, in this small group the results of arrhythmia assessment and of exercise ST segment analysis
provided no additional prognostic information. Of note, preliminary
results of our currently ongoing evaluation of patients studied within 3 weeks after infarction suggests that a combination of resting
radionuclide cineangiographic results and exercise ST segment findings may be particularly useful in selecting patients at high risk
for sudden death [13]. In our earlier study, exercise ejection fraction values also failed to modify prognoses based on resting ejection
fraction alone. However, the high sensitivity of exercise radionuclide cineangiography in detecting ischemia[14], and in identifying
patients with particularly severe left main and proximal 3 vessel
coronary stenoses[15] (and below), indicate that the exercise approach
may prove a valuable adjunct to assessment of resting ejection fraction alone in selecting patients at high risk for sudden death.
Clearly, further study is indicated in this area.

2. Electrophysiologic Testing

     While sudden death most often is associated with the development
of intractable ventricular arrhythmia, conventional means of arrhythmia detection appear to be of limited value in identifying patients
at highest risk. Thus, as noted above, the induction of arrhythmias
by exercise is a relatively constant indicator of prognosis; ambulatory electrocardiography during in hospital convalescence after infarction is, by itself, similarly deficient in identifying patients
likely to die suddenly[5,11,12,16]. However, the development of invasive programmed ventricular stimulation techniques, involving
catheter-mediated interposition of electrical stimuli to the ventricular myocardium during potentially vulnerable periods of the cardiac
cycle, holds promise for more precise identification of patients
with highest proclivity for the development of lethal arrhythmias.
Such testing now has been employed in a large group of patients
studied within one month after infarction, and, while the prognostic
value of such testing is not yet clear, preliminary results suggest
that electrophysiologic testing may be of greater predictive value
than exercise testing[17]. Moreover, if results in patients after
infarction are analogous to those of other high risk populations
that have undergone formal electrophysiologic testing, it is likely
that relatively precise determination of sudden death risk may be
expected from results of studies now underway[17,18].

Detection of patients at risk of sudden death, by itself, does not necessarily result in prevention of the lethal event. However, recent studies indicating the efficacy of beta adrenergic blockade in reducing sudden death in subsets of high risk patients during six months after infarction suggests that pharmacologic therapy currently may be available to markedly mitigate risk[19], thus providing additional impetus to attempts to identify patients most likely to benefit from such therapy.

B. PATIENTS WITH CHRONIC, STABLE ISCHEMIC DISEASE

Though sudden death risk, and overall mortality risk, is considerably lower in patients with clinically evident chronic, stable ischemic heart disease than in survivors of recent acute infarction (2 to 4% annual mortality in the former group vs 15% during the 6 to 12 months post infarction in the latter group), subgroups of patients at relatively higher risk can be identified. Thus, patients with severe stenosis ($\geqslant$75% luminal diameter narrowing) of the left main coronary artery, and patients with similarly severe stenosis of the proximal portions of all 3 major coronary arteries, manifest a 10 to 15% annual mortality risk from the time of discovery of their lesions; a high percentage of the resulting deaths are sudden. Most importantly, recent studies indicate that the fatal outcome in such patients can be significantly diminished by application of coronary artery bypass grafting therapy, thus indicating the practical clinical importance of detecting such patients from among the larger group of clinically indistinguishable patients with less severe ischemic disease.

Preliminary results of our studies in 250 patients with coronary artery occlusive disease and minimal to moderate ischemic symptoms suggests that radionuclide cineangiography during exercise may be of particular value in identifying patients with coronary artery lesions known to confer high short-term mortality risk. (Parenthetically, as a corollary to conclusions based on findings, our results are consistent with the concept that the functional severity of ischemic disease, as defined by radionuclide cineangiography, may be a more effective predictor of sudden death risk than is invasive arteriography; thus, mortality risk in patients with lesions not involving the left main artery or the proximal portion of all 3 major arteries, may be best defined with reference to the functional concommitant of the anatomic abnormality).

To assess the accuracy of radionuclide cineangiography in identifying patients with known high risk lesions, we compared the results of coronary arteriography with the magnitude of changes in left ventricular ejection fraction from rest to maximal, symptom limited supine bicycle exercise[15]. Radionuclide cineangiography was performed using the standard methods we have previously described[14,20,21]. Of our 250 patients, 10% had severe ($\geq$75% luminal diameter narrowing) left main coronary artery stenoses, and 18% had severe stenoses of the proximal portions of the 3 major coronary arteries. Our preliminary results indicate that, compared with resting values, left ventricular ejection fraction either rose during exercise, or fell <5% (the error of the method, defined as 2 standard deviations from the mean of repeated determinations from a single study) in 31% of patients[15]. None of these patients manifested severe left main coronary stenosis, and very few manifested proximal three vessel disease. In contrast, almost 40% of patients evidenced a marked ($\geq$10%) fall in left ventricular ejection fraction during exercise, and more than half of these patients manifested severe left main or proximal 3 vessel stenoses at coronary arteriography. Moreover, conversely, of the 70 patients with severe left main or proximal 3 vessel stenosis, almost three-fourths manifested $\geq$10% fall in ejection fraction during exercise, thus evidencing a ten-fold greater risk of such prognostically important lesions than was noted among the one third of our population with relatively mild ischemia involving <5% fall in ejection fraction during exercise[15].

II.   NON-ISCHEMIC HEART DISEASE

Patients with valvular heart disease and with cardiomyopathy also are known to be at relatively high risk of sudden death as compared with the general population.

While elucidation of the factors most predictive of sudden death in these patients has not been achieved, recent preliminary studies in our laboratories hold promise for identification of these risk factors in the relatively near future. For example, among patients with aortic regurgitation, while the incidence of sudden death prior to surgical valve replacement has declined considerably with the increasing availability of such surgery, it is known that sudden death may account for as much as 50% of the mortality observed within 5 years after therapeutic valve replacement. Patients at risk for such lethal events presumably may have suffered an irreversible biochemical derangement prior to operation which has resulted in intractible myocardial electrophysiological instability. Recently, we have found Lown Class III, IV or V ventricular arrhythmias during ambulatory electrocardiography in almost half our patients with isolated aortic regurgitation, and have observed that such abnormalities are closely associated with depression of left and right ventricular functional reserve determined at radionuclide cineangiography, as well as with depression of values of echocardiographic descriptors of left ventricular performance known to be predictive of post-operative survival[22]. These results suggest the possibility that correlation of mechanical and electrophysiological study, perhaps involving formal electrophysiological programmed stimualtion (see above), may be of particular value in identifying patients at risk

of sudden death after operation, for whom prophylactic antiarrhythmic
therapy might be of particular value.

III. CONCLUSION

In summary, prevention of sudden death in all patients at risk
has not yet been achieved. However, recently, major advances have
been reported in the detection of patients at highest risk of sudden
death. These developments should result in more effective application of currently available prophylactic therapy than is now prac-
ticed, and provides the opportunity for identification of group of
patients most appropriate for the pathophysiologic assessments necessary for development of increasingly effective therapy.

REFERENCES

1. D. H. Miller and J. S. Borer, Exercise testing early after myo-
    cardial infarction: Risks and benefits, Amer. J. Med. 72:
    427-438 (1982).
2. P. Theroux, D. D. Waters, C. Halphen, J. C. Debaisieux and
    H. F. Mizgala, Prognostic value of exercise testing soon after
    myocardial infarction, New Engl. J. Med. 301:341-345 (1979).
3. M. Ericsson, A. Granath, P. Ohlsen, T. Sodermark and U. Volpe,
    Arrhythmias and symptoms during treadmill testing three weeks
    after myocardial infarction in 100 patients, Brit. Heart J.
    35:787-790 (1973).
4. J. W. Smith, C. A. Dennis, A. Gassmann, J. A. Gaines, M. Staman,
    B. Phibbs and F. I. Marcus, Exercise testing three weeks after
    myocardial infarction, Chest 75:12-16 (1979).
5. R. A. Shulze, Jr., H. W. Strauss and B. Pitt, Sudden death in
    the year following myocardial infarction: Relation to ventri-
    cular premature contractions in the late hospital phase and
    left ventricular ejection fraction, Amer. J. Med. 62:192-199
    (1977).
6. M. Sami, H. Kramer and R. F. DeBusk, The prognostic significance
    of serial exercise testing after myocardial infarction,
    Circulation 60:1238-1246 (1979).
7. M. R. Starling, M. H. Crawford, G. T. Kennedy and R. A. O'Rourke,
    Exercise testing early after myocardial infarction Predictive
    value for subsequent unstable angina and death, Amer. J.
    Cardiol. 46:909-914 (1980).
8. K. J. Silverman, L. C. Becker, B. H. Bulkley et al., Value of
    early thallium - 201 scintigrapht for predicting mortality
    in patients with acute myocardial infarction, Circulation
    61:996-1003 (1980).
9. L. C. Becker, The early post infaction patient: Identification
    and aggressive intervention in high risk groups, Abstract,
    Proceedings of Conference on Sudden Coronary Death, New York
    Academy of Sciences, New York, p. 33 (1981).

10. A. Battler, R. Slutsky, J. Karliner, V. Froelicher, W. Ashburn
    and J. Ross, Jr., Left ventricular ejection fraction and first
    third ejection fraction early after acute myocardial infarc-
    tion: Value for predicting mortality and morbidity, Amer. J.
    Cardiol. 45:197-202 (1980).
11. R. Shulze, J. Rouleau, P. Rigo, S. Bowers, H. W. Strauss, P,H,B,
    Ventricular arrhythmias in the late hospital phase of acute
    myocardial infarction, Circulation 52:1006-11 (1975).
12. J. S. Borer, D. R. Rosing, R. H. Miller et al., Natural history
    of left ventricular function during one year after acute myo-
    cardial infarction: Comparison with clinical, electrocardio-
    graphic and biochemical determination, Amer. J. Cardiol. 46:
    1-12 (1980).
13. D. H. Miller, J. S. Borer, P. D. Kligfield, D. L. Hayes and
    E. Whitacre, Independence of exercise-induced electrocardio-
    graphic ischemia and left ventricular function in patients
    with recent myocardial infarction, Clin. Research 30 (in
    press) (1982).
14. J. S. Borer, K. Kent, S. L. Bacharach et al., Sensitivity,
    specificity and predictive accuracy of radionuclide cineangi-
    ography during exercise in patients with coronary artery
    disease, Circulation 60:572-580 (1979).
15. P. Phillips, J. S. Borer, J. Jacobstein, K. Plaucher, J. Carter,
    J. Moses, J. Goldstein, M. Collins and J. Fisher, Prognosti-
    cally critical coronary stenoses: Identification by radio-
    nuclide cineangiography, Amer. J. Cardiol. 49:991 (1982).
16. R. F. DeBusk, D. M. Davidson, N. Houston and J. Fitzgerald,
    Serial ambulatory electrocardiography and treadmill exercise
    testing after uncomplicated myocardial infarction, Amer. J.
    Cardiol. 45:547-554 (1980).
17. J. B. Uther, D. V. Cody, A. R. Denniss, D. A. Richards,
    P. A. Russell and A. A. Young, Ventricular programmed stimu-
    lation and exercise ST segment depression as predictors of
    post infarction mortality, Amer. J. Cardiol. 49:902 (1982).
18. C. Friedman, J. Moses, J. S. Borer, H. G. Goldberg, D. H. Miller
    and J. Fisher, Prognostic value of programmed ventricular
    stimulation early after infarction in high risk patients,
    Circulation (submitted).
19. Norwegian Multicenter Study Group, Timolol-induced reduction
    in mortality and reinfarction in patients surviving acute
    myocardial infarction, New Engl. J. Med. 304:801-807 (1981).
20. J. S. Borer, S. L. Bacharach, M. V. Green, K. M. Kent,
    S. E. Epstein and G. S. Johnston, Real-time radionuclide
    cineangiography in the non-invasive evaluation of global and
    regional left ventricular function at rest and during exercise
    in patients with coronary artery disease, New Engl. J. Med.
    297:839-844 (1977).
21. J. S. Borer, S. L. Bacharach, M. V. Green, K. M. Kent,
    G. S. Johnston and S. E. Epstein, Effect of nitroglycerin
    on exercise-induced abnormalities of left ventricular regional

function and ejection fraction in coronary artery disease:
Assessment by radionuclide cineangiography in symptomatic
and asymptomatic patients, Circulation 57:314-320 (1978).

22. C. Hochreiter, J. S. Borer, P. Kligfield, R. Devereux,
J. Jacobstein and M. Kase, Complex ventricular arrhythmias
in patients with valvular regurgitation: A potentially
important, clinically overlooked phenomenon?, Amer. J. Cardiol.
49:910 (1982).

THE INFLUENCE OF PSYCHOSOCIAL STRESSES ON SUDDEN DEATH

M. G. Marmot

Department of Medical Statistics & Epidemiology
The London School of Hygiene & Tropical Medicine
Keppel Street (Gower Street)
London WC1E 7HT

INTRODUCTION

John Hunter, apart from his other contributions to medical
science, has left a vivid description of his heart condition. His
angina pectoria was brought on by "agitation of the mind ... princi-
pally anxiety or anger ... the most tender passions of the mind did
not produce it". (Home 1794). His life was said to be in the hands
of any rascal who cared to provoke him. Apparently his sudden
death was brought on by a particularly irritating Hospital Board
Meeting. The post-mortem appearance of his coronary arteries were
described by his brother-in-law as being "in the state of bony
tubes" (Home 1794).

We are now in a position to confirm that some of these observations have general validity: sudden death is common in patients
with ischemic heart disease (including previously unrecognized
disease); and the majority of patients who die suddenly from presumed cardiac causes have evidence at autopsy of chronic severe
atherosclerosis and chronic ischemic heart disease.

Whether psycho-social factors can increase the risk of sudden
death has been more difficult to show. To examine this question,
we shall have to ask first if the predictors of sudden death differ
from those of myocardial infarction; and second if psycho-social
influences affect specifically the risk of sudden death as well as
having an effect on non-fatal coronary heart disease.

STATEMENT OF THE PROBLEM

In industrialized countries coronary atheroma is highly prevalent as are coronary risk factors: elevation of plasma cholesterol,
blood pressure and smoking. Yet most people with elevated levels
of these risk factors do not die prematurely (e.g. before age 70).
Among people at risk, it would be of prime importance to identify
extra factors leading to sudden death or to myocardial infarction.
If sudden death is but one complication of an acute ischemic episode,
then the search for factors specific for sudden death may be mis-
taken - the effort should be directed to finding factors that precipitate any acute ischemic attack. If on the other hand the pathology of sudden death is different from that of myocardial infarction, specific factors may be found. These may be psycho-social.

The evidence for sudden death as an entity separate from non-
fatal myocardial infarction will be reviewed below, first by considering pathological findings, then epidemiological studies of
predictors of sudden death, and the evidence for a specific prodrome
of sudden death. If, as seems the case, many of the predictors of
non-sudden cardiac death are similar to those of sudden cardiac
death, then psycho-social factors associated with one should be
associated with the other. These will be reviewed briefly, as will
the evidence for psycho-social influences specific to sudden death.

DEFINITION AND FREQUENCY

Sudden cardiac death can occur in patients with or without
previously diagnosed coronary heart disease. The word "sudden"
refers not to the time elapsed since the onset of the first symptom
of coronary heart disease but to the time from the first symptom
or sign of the terminal episode. In a patient with chronic symptomatic heart disease the terminal episode may be difficult to
define; and the duration of symptoms may be impossible to determine
in people who die unobserved.

Sudden death had been defined as occurring within 24 hours of
onset of symptoms (Paul & Schatz, 1971). Others have used 1 hour
as the dividing line (Moss, 1980). Friedman et al. (1973) separated
"instantaneous" (within 30 seconds) from other sudden deaths. Difficulty arises because sudden deaths do not appear to be a clearly
definable group. Figure 1 shows data from the Tower Hamlets (East
London) Coronary Heart Attack Register (Tunstall Pedoe, 1978), a
population study. The mortality is high in the first 15 minutes
after the onset of symptoms and continues to rise thereafter in
linear fashion, with elapsed time on a log scale, i.e., the rate
of increase of mortality slows progressively, with no clear break
point between "sudden" and "non-sudden".

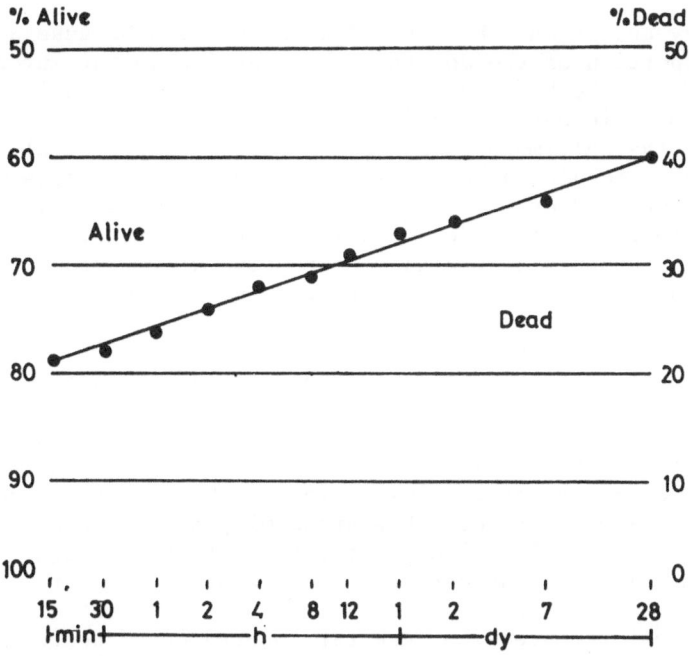

Fig. 1. Cumulative fatality rate against time in 878 coronary heart attacks in men and women in Tower Hamlets. (Tunstall Pedoe, 1978)

In his longitudinal study of middle-aged American men, Hinkle (1982) distinguishes "arrhythmic" death from death in circulatory failure, among men whose deaths were observed. All observed deaths that occurred in less than 5 minutes from the onset of symptoms were thought to be "arrhythmic"; 91% of deaths in less than 1 hour were arrhythmic; 85% of deaths in less than 24 hours were arrhythmic; whereas only 24% of deaths in more than 24 hours were arrhythmic. If these results are generally applicable, deaths occurring in less than 1 hour are likely to be a relatively homogeneous group of arrhythmic deaths.

Figure 1 also illustrates that sudden death is worthy of attention. Nearly 25% of heart attacks were fatal within 1 hour; and of all the fatalities within 28 days, approximately 60% occurred in the first hour. Sudden death, therefore, forms a major part of the mortality picture in middle-aged men.

IS SUDDEN DEATH DISTINCT PATHOLOGICALLY?

Autopsy series must be treated with caution, because so often
a large proportion of the deaths are not available for autopsy.
Nevertheless, there have been numerous reports of extensive coronary
atherosclerosis in patients dying a sudden cardiac death. For
example, in the Baltimore Sudden Death Study (Perper et al., 1975),
of the people who died a sudden cardiac death in 76% there was
⩾75% narrowing of 2 or more coronary arteries. In this respect they
differ little from non-sudden cardiac deaths. A major difference
is the high frequency of occlusive coronary artery thrombi (90%)
in the non-sudden deaths compared to the sudden (30%) (Moss, 1980).

This would point to an electrical event in a heart with extensive chronic disease as a possible cause of sudden death. This is
supported by the findings from Seattle on patients resuscitated
from a cardiac arrest outside hospital (Baum et al., 1974). Of
patients resuscitated from ventricular fibrillation, 50% showed no
evidence of acute myocardial infarction, although chronic coronary
heart disease was the rule. This group with no evidence of infarction had a substantially higher fatality over the subsequent two
years than patients whose cardiac arrest went on to acute infarction.

Others point to somewhat greater importance of acute myocardial
infarction. Hinkle suggests that 60% of "electrical" deaths are
caused by an acute ischemic episode. A report from Helsinki found
evidence of a recent myocardial infarction in 77% of persons dying
instantaneously and 73% of those dying within 2 hours of onset of
symptoms (Rissanen et al., 1978). The authors interpreted this as
showing that only a quarter of the deaths within 2 hours were due
to "primary arrhythmia" with no evidence of infarction - somewhat
lower than other estimates. It has been suggested that coronary
spasm may play a role in the aetiology of these arrhythmias (JAMA.,
1981).

ARE THERE UNIQUE PREDICTORS OF SUDDEN DEATHS?

Table 1 examines data recalculated from the 16 years follow-up
of the Framingham cohort (Shurtleff, 1970). Among younger men,
smoking, serum cholesterol and alcohol consumption are more strongly
related to sudden death than to non-sudden death or non-fatal myocardial infarction. This is not so for older men. Relative weight
is associated with sudden death in the older but not the younger.
In general, however, the risk factors for sudden death do not differ
greatly from those for non-sudden death or myocardial infarction.
This is the conclusion reached by Doyle et al. (1976) in their
analysis of data from Albany and Framingham. The longitudinal community study in Tecumseh also fails to find a specific profile for
people at risk of sudden death (Chiang et al., 1970).

Table 1. Risk gradients for men in the Framingham study for
          various forms of CHD, based on ratio of smoothed rates
          in the highest and lowest classes for each factor
          (calculated from Shurtleff, 1970)

| Factor | Risk Ratio (highest:lowest class) | | | | | |
| | 45-54 yr | | | 55-64 yr | | |
	Sudden death	Non-Sudden death	M.I.	Sudden death	Non-Sudden death	M.I.
Systolic BP	3.0	35.7	2.9	7.7	10.0	2.7
Serum Cholesterol	6.9	1.4	3.5	2.0	6.5	1.9
Cigarettes	6.0	1.3	2.8	1.5	1.9	1.7
Alcohol	1.7	0.3	0.5	0.7	0.4	0.9
Relative weight	1.2	0.9	2.3	6.3	1.1	1.5
Blood sugar	1.6	0.5	3.3	1.4	1.1	2.0

Against this, Hinkle (1982b) finds heavy alcohol consumption
and smoking to be more strongly associated with risk of arrhythmic
death than death in circulatory failure, but confirms that prior
evidence of ischemic heart disease is strongly associated with
sudden death.

SUDDEN DEATH AFTER MYOCARDIAL INFARCTION

There is no strong and consistent evidence for a specific risk
profile distinguishing sudden from non-sudden death or non-fatal
recurrence after a myocardial infarction. Moss (1979) found similar
proportions of smokers, diabetics and hypertensives among post-MI
sudden deaths as among non-sudden. By contrast, Ruberman et al.,
(1975) did find complex ventricular ectopic beats to be more closely
associated with sudden than non-sudden cardiac deaths.

If the mechanism of sudden cardiac death after a myocardial
infarction were different from that of non-sudden death, one might
expect the response to therapy to be different. Recent trials of
β-blockers show a reduction in mortality after a myocardial infarction (Norwegian Multicentre Study Group 1981; B.H.A.T., 1981;
Hjalmarsan, 1981). A distinction between sudden and non-sudden death
was looked for only in the Norwegian Timolol study which showed a
reduction in both, altogether later evidence apparently shows the
effect on sudden death to be greater (Sleight, 1982).

The first Anturane reinfarction trial reported a specific reduction in sudden cardiac death (A.R.T., 1980), but this was not
confirmed by the subsequent Anturan Reinfarction Italian Study (1982).
Similarly, the trials of aspirin in secondary prevention of an

myocardial infarction have not shown a consistent effect on sudden
death specifically (Genton, 1980).

A SPECIFIC PRODROME OF SUDDEN DEATH?

From his clinical observations, Nixon (1976) has described a
stage of depression and exhaustion that precedes a myocardial infarction. He relates this prodrome to the individuals inability to
continue to cope with the demands upon him. Similarly, Appels (1980)
has shown that vital exhaustion and depression are significantly
associated with the development of myocardial infarction. It is not
clear from this work if there are different prodromes for sudden
death and non-fatal myocardial infarction.

The question of a specific prodrome is relevant here as it may
help (a) to distinguish precipitants for sudden death from those for
myocardial infarction and (b) to predict those at immediate risk of
sudden death, on the assumption that preventive action may be effective.

Based on retrospective studies (Table 2), chest pain and
dyspnoea appeared to have been reported with equal or greater frequency in non-fatal myocardial infarction than in sudden death.
Kinlen's Oxford Study shows an excess of 'fatigue' which might correspond to Appels 'vital exhaustion'. These comparisons between surviving and non-surviving patients are difficult to interpret,
because in the case of survivors the respondents are usually the
patients themselves, and in the sudden deaths, relatives of the
deceased. One study of surviving patients, found symptoms to be
over-reported by relatives compared to the patients themselves
(Gillum, 1976).

Thus there does not appear to be a prodrome specific for sudden
death, although Appels' study of patients thought to have imminent
myocardial infarction and controls does suggest that an acute coronary event, fatal or non-fatal, is preceded by a syndrome of exhaustion and depression. One explanation for this is that psycho-
social factors lead both to exhaustion and depression and to an acute
coronary attack. An alternative explanation is that worsening
cardiovascular pathology leads both to symptoms and to an acute
coronary attack. An attempt could be made to distinguish between
these by studying the relation between cardiac function and symptoms
in the study of imminent myocardial infarction.

PSYCHO-SOCIAL INFLUENCES ON CORONARY HEART DISEASE

These pathological, epidemiological and clinical studies suggest
that factors that influence risk of developing coronary heart disease

Table 2. Comparative Frequency of Recent Symptoms in Cases of Non-fatal Myocardial Infarction and Sudden Death

Authors	No. of Patients		Chest Pain		Fatigue		Dyspnoea	
	M.I.	S.D.	M.I.	S.D.	M.I.	S.D.	M.I.	S.D.
Simon et al., 1972, 1973	160	138	48%	22%	27%	26%	25%	25%
Kinlen, 1973	194	140	74%	69%	20%	41%	19%	21%
Feinleib et al., 1975	73	19	70%	32%	68%	68%	26%	21%

will also increase the risk of sudden death. This is likely to
extend to psycho-social factors. We should therefore consider,
(i) factors that increase risk of coronary heart disease without
necessarily being associated specifically with sudden death; and
(ii) the evidence for factors that precipitate sudden death specifically.

     Jenkins (1982) has lately divided psycho-social factors that
influence the risk of coronary heart disease into four categories:
socio-economic disadvantage, sustained disturbing emotions, Type A
behavior and psychological overload. Although there is overlap
between them, they serve as useful categories for this brief review.
They also provide a way of discussing psycho-social factors without
having to face the ambiguities of the word "stress".

Socio-economic Disadvantage

     In contrast to earlier part of this century, in many industri-
alized countries, people in lower social classes now have a higher
mortality rate from coronary heart disease than those in higher
classes (Marmot 1978a). In the "Whitehall" study of civil servants
(Figure 2), for example, men in the lowest grade of employment had
more than three fold higher mortality from CHD than men in the
highest grade (Marmot et al., 1978b). In this case, "social" factors
are partly 'biological/behavioral': higher rates of smoking, more
obesity, less physical activity in lower grades; but they may also
be 'psycho-social': less social support, less job satisfaction in
lower grades (Marmot, 1982). The change in social class distribution
of heart disease, from higher in upper classes to higher in lower

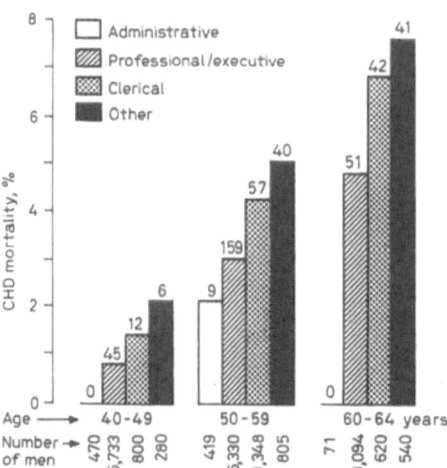

Fig. 2.   Coronary heart disease mortality in 7½ years by age and
          grade in the British Civil Service (Marmot, 1978).

classes, correlates with a change in smoking. It is more difficult
to chart a change in psycho-social factors.

In at least three studies, in Newcastle-upon-Tyne (Myers &
Dewar, 1975), New Zealand (Fraser, 1978), and New York (Weinblatt
et al., 1978) sudden death was more common in men of lower social
class and/or lower education.

Sustained Disturbing Emotions

In his review Jenkins (1982) lists a series of studies that
show anxiety, depression and neuroticism to be related to angina
pectoris but not to risk of myocardial infarction. It is possible
that this reflects not a relationship of emotions to cardiac pathology, but a relation between two kinds of symptoms complaints;
neuroticism and chest pain. However, in several of the longitudinal
studies cited by Jenkins, 'disturbing emotions' were related to
cardiac death. It remains a possibility that 'disturbing emotions'
are the result of some aspects of the body's functioning which also
increases the risk of coronary heart disease.

Type A Behavior

The coronary-prone behavior pattern described by Friedman and
Rosenman has been shown in several studies to be related to coronary
heart disease (Rosenman and Chesney, 1980). This relationship does
not hold equally in all populations. For example, in Framingham,
Type A was more strongly related to coronary heart disease in white
than in blue-collar workers (Haynes et al., 1980). In the Whitehall
study, Type A was more common among upper grades, whereas coronary
heart disease was more common in lower grades (Marmot, 1982). This
may reflect social and cultural differences in ways of behaving in
response to stress, but also in ways of answering questions in
interviews designed to assess behavior pattern.

Two important questions concerning Type A behavior: how does
it arise? And how does it exert its effects? It has always been
suggested by Friedman & Rosenman that it is the reaction of a predisposed individual to a demanding, intense, over-stimulating environment.

There are now studies using coronary angiography that show
Type A individuals to have more evidence of coronary atheroma than
Type B's (Zysanski et al., 1975). There is no general agreement on
this and atypical findings could result from bias in selection of
patients for angiography. Nevertheless, it is possible that Type
A exerts its effect on the chronic development of atherosclerosis.
In one postmortem study, using interviews of relatives, Friedman

et al. (1973) showed Type A to be related both to instantaneous
(presumably arrhythmic death) and to sudden death (< 24 hours). This
relationship could be indirect - the result of the association
between Type A and chronic coronary heart disease, as the hearts of
the persons dying instantaneously showed gross evidence of chronic
pathology.

Psychological Overload at Work

In their study of building workers in Stockholm, Theorell et
al. (1977) found that excessive work load increased the risk of
myocardial infarction. More recently, Karasek et al. (1981) identified two characteristics of jobs: decision latitude (roughly, inde-
pendence in making job decisions) and job demands. People in jobs
with high demands and low decision latitude were at excess risk of
cardiovascular disease. There have been other studies showing a
relationship of job characteristics to coronary heart disease
(Jenkins, 1982).

The studies summarized under these four headings lead to the
conclusion that psycho-social influences do increase the risks of
cardiac disease and this is likely to increase the risk of sudden
death as well as other manifestations of coronary heart disease.

SPECIFIC PSYCHO-SOCIAL INFLUENCES ON SUDDEN DEATH

There are anecdotal reports of sudden deaths occurring during
circumstances such as exercise, or psychological arousal (as in John
Hunter's case), or sexual intercourse. Such reports lack controls,
or at least the notion of a control distribution of events in time.
The question should be: do stressful events occur in the period
preceding a sudden death more often than one would expect by chance?
When more systematic comparison is made with controls who have not
died, one must still consider the possibility, referred to above,
that relatives of cases faced with the catastrophe of death, may
subconsciously bias their reports of circumstances preceding death.

Siltanen (1978) reported that 19% of sudden death cases in
Helsinki experienced marked stress in the 12 hours preceding death.
This appears to be greater than one would expect by chance, although
possibly the result of biased reporting. They did not report the
frequency of marked stress in non-fatal attacks.

In Myers & Dewar's (1975) study of 100 sudden coronary deaths
referred to the coroner, in Newcastle-upon-Tyne, 23% were reported
to have experienced 'acute stress' in the 30 minutes prior to death,
and 40% in the 24 hours. By contrast in 100 non-fatal myocardial
infarctions, the figures were 8% in the preceding 30 minutes and 24%
in 24 hours (Table 3). There was no difference in chronic stress.

Table 3. Percent reporting psychological stress in relation
          to sudden death and onset of non-fatal myocardial
          infarction

	Sudden Death N = 100	Non-fatal Myocardial Infarction N = 100
1. Acute stress		
a) In preceding 30 mins		
None/minimal	77%	92%
Moderate	18%	5%
Severe	5%	3%
b) In preceding 24 hrs		
None/minimal	60%	76%
Moderate	30%	18%
Severe	10%	6%
2. Chronic stress		
None/minimal	72%	62%
Moderate	16%	23%
Severe	12%	15%

Myers & Dewar (1975)

What constitutes "stress", an important issue, appears to be a
matter of joint judgement between the investigator and the respondent
in these studies. In a separate retrospective study from Helsinki,
Siltanen and colleagues (1978) used a standard questionnaire, the
Schedule of Recent Experience (SRE). This includes a comprehensive
list of potentially stressful life changes, weighted according to
severity, which yields a total 'life-change' score (Rahe et al.,
1974). Comparing the 6 months preceding a coronary attack with the
corresponding 6 months one year earlier, there was a greater build
up of 'life-changes' in men dying a sudden death than in survivors
of a myocardial infarction (Table 4). This association was not seen
in women.

In a prospective study, Theorell et al. (1977) found no association between life-events scores and subsequent coronary deaths.
Relationships did emerge when they looked, not at total life changes,
but at specific changes - in particular increased work load. This
was associated with increased risk of non-fatal myocardial infarction
and not sudden death. The general negative findings from this prospective enquiry into life-events and sudden death does increase
the likelihood that the positive results based on case-control
studies may be influenced by biased reporting of events. Against

Table 4. 6 Months "Life-change" Scores and % Increases 6 Months
          preceding the Heart Attack compared with 12 Months earlier

	N	6 Month life change scores in period preceding attack:		% Increases
		12 - 18 months	0-6 months	
No recent illnesses				
Survivors	166	23	39	69
Sudden deaths	61	21	51	143
Delayed deaths	35	35	42	20
With recent illnesses				
Survivors	133	52	74	42
Sudden deaths	65	43	77	79
Delayed deaths	49	56	96	71

Siltanen (1978)

this, however, is the suggestion from Myers & Dewar's study that the
stress must take place within 24 hours of the attack. In this case
it is likely to be missed, when the follow-up period from events to
fatal attack is 1-2 years.

An intriguing finding has been that of Weinblatt et al. (1978)
in their follow-up of survivors of a myocardial infarction. In the
presence of complex ventricular premature beats on 1-hour ECG monitoring, low education was associated with a more than 3-fold increase
in sudden death - occurring within minutes (Figure 3). There was
no association with non-fatal recurrence of myocardial infarction.
They could not explain this association by any other known predictors
of survival after a myocardial infarction. They speculated that
among those of less education there may have been a higher rate of
occurrence of stressful circumstances leading to ventricular fibrillation. A relationship between low social class and sudden coronary
death was also found by Myers & Dewar (1975).

Ventricular Arrhythmias and Sudden Death

Lown (1979) and colleagues have been in the forefront in putting
the case that ventricular arrhythmias may be the cause of sudden
death. Lown emphasizes the two aspects: a chronic predisposition to
electrical instability of the ischemic myocardium and an acute precipitating factor. In animal experiments, Lown and colleagues found
that aversive psychological stimuli could induce ventricular fibrillation in dogs with occlusion of the left anterior descending coronary artery. This is likely to be mediated by sympathetic arousal,
as pharmacological β-blockade abolished this effect (Lown, 1978).

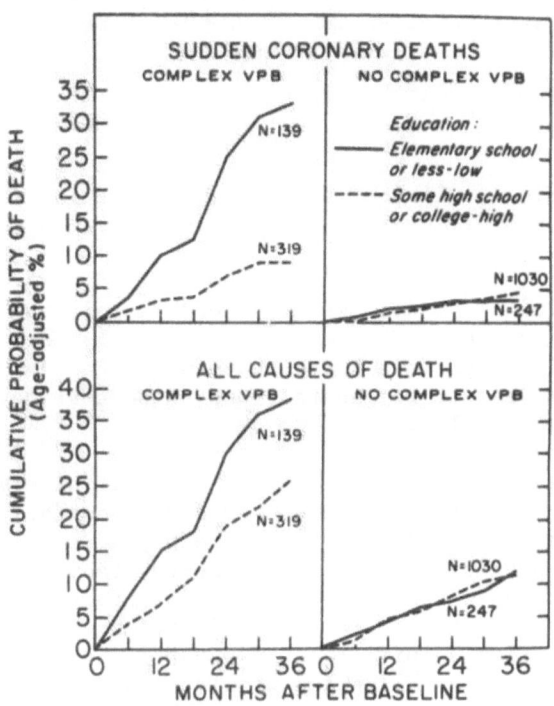

Fig. 3. Mortality over three years among men with prior myocardial
         infarction, according to educational level and presence of
         complex ventricular premature beats (VPB) in the base-line
         monitoring hour (Weinblatt et al., 1978).

        The possible relevance to humans was seen in a study of 117
patients referred for anti-arrhythmic management, 21% had an acute
emotional disturbance in the 24 hours preceding arrhythmias, and 18
patients (15%) had two or more documented arrhythmic episodes preceded by psychological disturbances (Reich et al., 1981). In this
study the evidence for association is not that the patients with

arrhythmias had more emotional disturbance than controls, but that, within the patient group, there was a clustering in time of emotional disturbance and arrhythmias. This is based on the assumption that the clustering did not result from bias in patients' reports. Consistent with Reich's findings, Hinkle (1982a) reported that 34% of the arrhythmic deaths observed in his sample occurred during activities associated with sympathetic arousal.

CONCLUSION

The pathological, clinical and epidemiological features of sudden death are broadly similar to those of other forms of acute heart attacks.

The prediction of sudden death is very largely the prediction of myocardial infarction. The two kinds of psycho-social influence on sudden death are therefore those that influence the development of coronary heart disease in general, and those that may precipitate sudden death in particular.

It has proved far more difficult to show definitely that there are specific psycho-social influences on sudden death. There are methodological reasons for this. If these psycho-social stresses act as immediate precipitants of sudden death, they will be missed in longitudinal studies that assess base-line characteristics and then follow subjects over several years. Attempts to show association in retrospective enquiries of cases and controls are subject to the worry that the event (death) has influenced assessment of whether stress preceded the attack. Nevertheless, these retrospective studies have suggested that stress may act as an immediate precipitant of sudden death. The finding that ventricular arrhythmias may also be precipitated by emotional stress is consistent with this.

One may speculate that there are two groups of sudden death: 30-60% in whom the ventricular arrhythmia is precipitated by an acute ischemic attack, which resulted in classic myocardial infarction, or would have, had not death supervened. In the other group, the arrhythmia may be the result of electrical instability in a chronically ischemic myocardium, and perhaps triggered by sympathetic arousal.

These results suggest that some progress may be made in the prevention of sudden death. After a myocardial infarction when the level of risk is high, drug prophylaxis is warranted. The encouraging results from the use of β-blockers may result from preventing the effects of sympathetic arousal. The comparable reduction in myocardial infarction (if genuine) may derive from other effects of sympathetic arousal, e.g., on thrombosis.

In average-risk people who have not suffered a myocardial infarction, too many people would have to take drugs for too little gain for drug therapy to be useful. The prevention of sudden death is the prevention of coronary heart disease. There may however be an intermediate group with diagnosed ventricular arrhythmias in whom some form of stress management may avoid the precipitants of sudden death.

REFERENCES

Anturan Reinfarction Italian Study, 1982, Sulphinpyrazone in post-myocardial infarction, Lancet, i:237-242.

Anturane Reinfarction Trial Research Group, 1980, Sulfinpyrazone in the prevention of sudden death after myocardial infarction, New Eng. J. Med., 302:250-256.

Appels, A., 1980, Psychological prodromata of myocardial infarction and sudden death, Psychother. Psychosom., 34:187-195.

β-Blockers heart attack Study Group, 1981, The β-Blocker Heart Attack Trial, J. Am. Med. Ass., 246:2073-4.

Baum, R. S., Alvarez, H. III, Cobb, L. A., 1974, Survival after resuscitation from out-of-hospital ventricular fibrillation, Circulation, 50:1231-35.

Chiang, B., Perlman, H.V., Fulton, M., Ostrander, L. D., Epstein, F. H., 1970, Predisposing factors in sudden cardiac death in Tecumseh, Michigan: a prospective study, Circulation, 41:31-37.

Doyle, J. T., Kennel, W. B., McNamara, P. M., Quickenton, R., Gordon, T., 1976, Factors related to suddenness of death from coronary disease: combined Albany-Framingham studies, Am. J. Cardiol., 37:1073-78.

Feinleib, M., Simon, A. B., Gillum, R. F., Margolis, J. R., 1975, Prodromal symptoms and signs of sudden death, Circulation, 51 and 51 Suppl. III:155-159.

Fraser, G. E., 1978, Sudden death in Auckland, Aust. N.Z. J. Med., 8:490-499.

Friedman, M., Manwaring, J. H., Rosenman, R. H., Donlon, G., Ortega, P., Grube, S. M., 1973, Instantaneous and sudden deaths: Clinical and pathological differentiation in coronary artery disease, J. Am. Med. Assoc., 225:1319-1328.

Genton, E., 1980, A perspective on Platclet-suppressant drug treatment in coronary artery and cerebrovascular disease, Circulation, 62:Suppl. V:111-121.

Gillum, R. F., Feinleib, M., Margolis, J. R., Fabsitz, R. R., Brasch, R. C., 1976, Community surveillance for cardiovascular disease: the Framingham disease survey, J. Chron. Dis., 29:289-299.

Haynes, S. G., Feinleib, M., 1982, Type A Behaviour and the incidence of coronary heart disease in the Framingham Heart Study in H. Denolin Ed. Psychological problem before and after myocardial infarction, Adv. Cardiol., 29:85-95, Karger, Basel.

Hinkle, L. E., 1982a, The immediate antecedents of sudden deaths, Acta Medica. Scandinavica, in press.

Hinkle, L. E., 1982b, Short term risk factors for sudden death, Bulletin of the New York Academy of Sciences, in press 1982.

Hjalmarson, Elmfeldt, D., Horlitz, J., Holmberg, S., Malek, I., Hyberg, G., Ryden, L., Swedberg, J., Vedin, A., Waagstein, F., Waldenstrom, A., Waldenstron, J., Wedel, H., Wilhelmsen, L., Wilhelmsson, C., 1980, Effect on mortality of metoprolol in acute myocardial infarction, Lancet, ii:823-827.

Home, E., 1794, A short account of the author's life by his brother-in-law Everard Home, 1941, reproduced by Willius, F.A. and Keys, T.E., Cardiac Classics, St. Louis, C.V., Mosby.

JAMA., 1981, Ventricular arrhythmias may not be primary cause of sudden death, J. Am. Med. Ass., 246:581-589.

Jenkins, C. D., 1982, Psycho-social Risk Factors for Coronary Heart Disease, Acta. Med. Scand., (Suppl.), in press.

Karasek, R. Baker, D., Marxer, F., Ahlbom, A., Theorell, T., 1981, Job decision latitude, job demands, and cardiovascular disease: a prospective study of Swedish men, Amer. J. Publ. Hlth., 71: 694-705.

Lown, B., 1979, Sudden Cardiac Death - 1978, Circulation, 60:1593-1599.

Marmot, M. G., 1982, Socio-economic and cultural factors in ischemic heart disease in H. Denolin Ed. Psychological problems before and after myocardial infarction, Adv. Cardiol., 29:68-76, Karger, Basel.

Marmot, M. G., Adelstein, A. M., Robinson, N., Rose, G. A., 1978a, Changing social-class distribution of heart disease, Brit. Med. J., 2:1109-1112.

Marmot, M. G., Rose, G. A., Shipley, M., Hamilton, P. J., 1978b, Employment grade and coronary heart disease in British Civil Servants, J. Epid. & Comm. Hlth., 32:244-249.

Moss, A. J., 1980, Prediction and Prevention of Sudden Cardiac Death, Ann. Rev. Med., 31:1-14.

Moss, A. J., Davis, H. T., Decamilla, J. D., Bayer, L. W., 1979, Ventricular ectopic deaths and their relation to sudden and non-sudden cardiac death after myocardial infarction, Circulation, 60:998-1003.

Myers, A., Dewar, H. A., 1975, Circumstances attending 100 sudden deaths from coronary artery disease with coroner's necropsies, Brit. Heart J., 37:1133-1143.

The Norwegian Multicentre Study Group, 1981, Timolol-induced reduction in mortality and reinfarction in patients surviving acute myocardial infarction, New Eng. J. Med., 304:801-807.

Nixon, P. G. F., 1976, The Human Function Curve, The Practitioner, 217:765-769 and 935-944.

Paul, O., Schatz, M., 1971, On Sudden death, Circulation, 43:7-10.

Perper, J. A., Kuller, L. H., Cooper, M., 1975, Arteriosclerosis of coronary arteries in sudden, unexpected deaths, Circulation, 51 and 52 Suppl. III:27-33.

Rahe, R. H., Romo, M., Bennett, L., Siltanen, P., 1974, Recent life changes, myocardial infarction and abrupt coronary death: studies in Helsinki, Arch. Intern. Med., 133:221-227.

Reich, P., DeSilva, R., Lown, B., Murawski, B. J., 1981, Acute psychological disturbances preceding life-threatening ventricular arrhythmias, J. Am. Med. Ass., 246:233-235.

Rissanen, V., Romo, M., Siltanen, P., 1978, Prehospital sudden death from ischemic heart disease: a post-mortem study, Brit. Heart J., 40:1025-33.

Rosenman, R. H., Chesney, M. A., 1980, The relationship of Type A behavior pattern to coronary heart disease, Activitas Nervosa Superior, 22:1-45.

Ruberman, W., Weinblatt, F., Frank, C. W., Goldberg, J. D., Shapiro, S., Feldman, C. L., 1975, Ventricular premature beats and mortality of men with coronary heart disease, Circulation, 52: Suppl. III:199-201.

Shurtleff, D., 1970, The Framingham Study: An epidemiological investigation of cardiovascular disease (ed. Kannel, W. B. and Gordon, T.). Section 26: Some characteristics related to the incidence of cardiovascular disease and death: The Framingham Study 16 year follow-up, U.S. Govt. Printing Office, Washington D. C.

Sleight, P., 1982, B-Blockade after myocardial infarction: Current Medical Literature, Cardiovascular Medicine, Royal Social of Medicine, 1:61-63.

Siltanen, P., 1978, Life changes and sudden coronary death, Adv. Cardiol., 25:47-50, Karger, Basel.

Simon, A. B., Alonzo, A.A., 1973, Sudden death in non-hospitalized patients, Arch. Int. Med., 132:163.

Simon, A. B., Feinleib, M., Thompson, H. K., 1972, Components of delay in the pre-hospital phase of acute myocardial infarction, Am.J. Cardiol., 30:476.

Theorell, T. and Floderus-Myrhed, B., 1977, Workload and myocardial infarction: a prospective psycho-social analysis, Int. J. Epidemiol., 6:17.

Tunstall Pedoe, H., 1978, Uses of coronary Heart Attack Registers, Brit. Heart J., 60:510-515.

Weinblatt, E., Ruberman, W., Goldberg, J. D., Frank, C. W., Shapiro, S., Choudhary, B. S., 1978, Relation of education to sudden death after myocardial infarction, New Eng. J. Med., 299:60-65.

Zysanski, S. J., Jenkins, C. D., Ryan, T. J., Flessas, A. et al., 1976, Psychological correlates of coronary angiographic findings, Arch. Intern. Med., 136:1234-1237.

PREVENTION OF SUDDEN DEATH:

ROLE OF ANTIARRHYTHMIC THERAPY

David H. Spodick

University of Massachusetts, Medical School
Division of Cardiology
St. Vincent Hospital

SCOPE OF THE SUDDEN DEATH PROBLEM IN THE UNITED STATES

Currently it is estimated there are about 600,000 cardiac deaths per year in the United States[1]. Of these 400,000 are relatively sudden, i.e. unexpected and rapid, amounting to 1,200 per day and therefore almost 1 per minute. Half of these "sudden" deaths occur within one hour from the onset of symptoms and many of these are truly sudden, i.e. instantaneous. This is notably a "disease" of men with up to 4 times the incidence in women. Indeed, in young and middle-aged men it is the leading cause of death - amounting to about 1/3 of deaths.

THE SUDDEN CARDIAC DEATH "SYNDROME"

If there is a stereotype for sudden death it is the occurrence, primarily in men, of instantaneous or rapid cessation of circulation in an individual who had, or subsequently is found to have, heart disease. In 3/4 of the cases this is ischemic although acute coronary thrombosis is found in only about 5% and acute myocardial infarction is identifiable in less than 20%. At least temporary myocardial electrical instability is certain to be present and recurrent or permanent instability is foreshadowed in many patients because of an annual recurrence rate averaging about 30%[2].

Etiology of Sudden Death

Sudden death can occur in any kind of heart disease, particularly coronary disease, already mentioned, with obstruction occurring in a minority of cases. Spasm is inferred for some others, but the mechanism is uncertain. Patients with both dilated (congestive) cardiomyopathy and hypertrophic cardiomyopathy are always in danger of sudden death[3]. Patients with mitral and/or tricuspid prolapse - conditions frequently associated with arrhythmias - occassionally die suddenly. Congenital etiologies include the Wolf-Parkinson-White and other pre-excitation syndromes and various complex congenital lesions, including post-surgical status (e.g. Tetralogy of Fallot). Acquired prolonged QT interval, for whatever reason, including medication (notably drugs used for antiarrhythmic effects) is associated with important and often fatal arrhythmias. Advanced atrioventricular block and the sick sinus syndrome - most patients with which are not ischemic - produce sudden death either due to ventricular asystole, lack of an escape rhythm during sinus asystole, and ventricular fibrillation (the last being the principal immediate mechanism of sudden death under most circumstances). Patients with bundle branch block are also in increased danger of sudden death, particularly those from hospital populations; ambulatory patients with bundle branch block seem to have less risk; severity of underlying disease is probably responsible. Finally, central nervous system disease with or without heart disease can predispose to sudden death, presumably via neural effects on cardiac repolarization causing susceptibility to malignant ventricular arrhythmias. Patients with acute myocardial infarctions have a variable risk for myocardial rupture, usually heralded by electromechanical dissociation - an important cause of in-hospital sudden death.

Immediate Cause of Sudden Death: Rhythm Disturbance

The immediate mechanism of sudden cessation of the circulation is apparent from the relatively few but increasing number of patients whose cardiac activity has been recorded at death on Holter monitors and from those who have been monitored by emergency medical personnel either at the place where they collapsed or in emergency rooms - confirming the primacy of ventricular fibrillation (VF) as the common final mechanism. Goldstein (1982)[4] noted cardiac arrest arrhythmias recorded initially by emergency medical squads. Among 104 patients, 87 had VF, six ventricular tachycardia (VT), eight asystole, and three complete atrioventricular block. VF did not develop terminally in relatively few. Among those with VF, 38 were associated with acute myocardial infarct (AMI), 37 with an ischemic event that was not infarction and 12 occurred as a primary arrhythmic event (i.e. no other acute cardiac pathologic change identified). The same author showed that almost 80% of patients identified as victims of sudden death had some history of heart disease; a history of previous AMI and of digitalis therapy characterized the majority of those with a primary arrhythmic event. (In this group also, cardiac arrest as the first cardiac event was uncommon). Arrest as first cardiac event characterized about 1/3 of those with AMI and half as many with ischemic events not producing infarction.

The importance of ischemic disease, with or without acute structural changes is apparent in the work of Lown[1] who showed that the mean VF threshold fell from 56 to 1.6 milliamperes and the vulnerable period for VF increased from 14 to 76 msec during experimental coronary occlusion with very incomplete restitution after release.

Hinkle and Thaler[5] compared arrhythmic deaths to deaths in circulatory failure on the basis of time of death following onset of symptoms. As might be predicted, almost half the arrhythmic deaths but none of the deaths due to circulatory failure occurred in less than 5 minutes. Thereafter, there was a reciprocal relation - deaths due to arrhythmia declined drastically to approximately 10% in the period between 24 hours and one week, while deaths in circulatory failure rose to 22%. In deaths occurring later than one week, there were 11% in the arrhythmic group and almost 60% for the circulatory failure group.

Pre-Hospital Cardiac Arrest

Out of 352 cardiac arrests[6], 220 were due to VF; 108 patients had bradyarrhythmias or asystole, and 24 had VT. Only among those with VT was a substantial number (21) admitted to the hospital alive. But only 87 of the 220 with VF, and only nine of those with bradyarrhythmias or asystole, reached the hospital alive. After reaching the hospital, however, the latter group had no further deaths. In contrast, there were five more deaths in the VT group and 36 in the VF group. Thus, of 220 ventricular fibrillations, 133 died before reaching the hospital and 36 died in the hospital with only 51 discharged alive. This accounts for the majority of out-of-hospital sudden deaths. (The prognosis was better in the few patients who had early resuscitation initiated by bystanders or rescue personnel)[6].

Adgey[7] noted that increasing heart rate occurring from the beginning of out-of-hospital monitoring into the period of movement to hospital was a distinct harbinger for subsequent VF. Almost 80% of his patients had late cycle VEBs but over 50% had R-on-T VEBs and almost 1/3 had consecutive VEBs. No extrasystolic activity was seen in 12%.

In-Hospital Sudden Death

It is common experience during early acute myocardial infarction - the condition most likely to be well monitored - that primary

ventricular fibrillation follows runs of ventricular ectopic beats
(though it may also occur without warning), while late ventricular
fibrillation follows ventricular tachycardia with high heart rate.
Other differences between early and late arrhythmia are seen in the
first 72 hours. Although early acute infarction patients studied in
the electrophysiology laboratory often have more resistant arrhythmias
these early arrhythmias do not predict sudden death. Thus, sudden
death mechanisms may resemble those of late (after 72-hr) ventricular
fibrillation in patients with acute myocardial infarction.

Myerburg[8] reported 38 patients with complex ventricular ectopic
activity, among whom ten had recurrent cardiac arrest due to VF within
72 hours of the initial event; 28 did not. The characteristics of
the arrhythmia's response to therapy, ventricular function and extent
of underlying heart disease did not separate those with from those
without recurrent cardiac arrest. Among these 38 patients, however,
12 had AV block, intra-ventricular conduction disturbances, or both,
either at admission or during the first 24 to 48 hours. Nine of those
12 were among the ten recurrent cardiac arrest victims. These patients
were from a group of 53 resuscitated and hospitalized pre-hospital
cardiac arrest victims; of these, 11 were never hemodynamically stabilized and died early in circulatory failure. Of the remaining 42
patients, 38 had complex ventricular arrhythmias and 4 had infrequent
unifocal VEBs.

Goldstein[4] identified a post-resuscitation high risk group for
sudden death, based on the following characteristics: 1) primary
arrhythmic event (with or without left ventricular dysfunction), 2)
history of digitalis therapy, 3) elevated BUN, 4) pulmonary congestion
in the hospital. In this high risk group, survival for one year was
only 71%. falling to 55% at two years. It is noteworthy that patients
with an acute myocardial infarction associated with the arrest had
a better prognosis. The mode of subsequent death in the high risk
group was sudden in 41% and due to congestive failure in 47%. Of the
remaining (low risk) group, 61% died and 25% had congestive heart
failure during the period of observation.

"Noncardiac" risk factors contributing to the possibility of
sudden death, include neurologic and psychologic factors elegantly
evaluated by Lown[1],[2]. Neurologic influences include complex interactions between the effects of the stellate ganglia, the central
nervous system, variations in blood pressure (and consequently sympathetic tone) and the effect of sleep as well as the modulating effect
of vagal stimulation. Psychologic stress also can play a role, the
type seeming not to be critical. Laboratory induction of malignant
arrhythmias in animals is facilitated by mental stress conditioning.
Thus, dogs kept in a nonaversive cage environment have a higher
ventricular fibrillation threshold than those in an aversive sling
environment, to which they have been conditioned[2].

MECHANISMS OF SUDDEN DEATH

It is apparent that the immediate precipitation of instantaneous death is usually the presence of ventricular fibrillation which may follow other ventricular ectopic activity, asystole and bradyarrhythmias, or electromechanical dissociation. The tissue mechanisms for ventricular ectopic activity are essentially two: re-entry and depolarizing after-potentials, providing clues for designing appropriate prevention and therapy. At the same time, pathogenic factors resulting in electrical instability must be considered. Some of these have already been touched on, but they may be listed as (1) autonomic, (2) central nervous, (3) metabolic, (4) mechanical - including dyskinesis and AV valve prolapse, and (5) coronary - including ischemia and left ventricular dysfunction, with heterogeneous tissue injury predisposing to micro- or macro-reentry in the myocardium, conducting tissues and cardiac nerves.

RELATIONSHIP OF ARRHYTHMIAS TO SUDDEN DEATH: PROBLEMS

It should be emphasized that the precise role of arrhythmias other than ventricular fibrillation is complex owing to problems involving patient factors and the ventricular arrhythmias themselves. Patient factors include: (1) among groups who have been the subjects of formal studies, sudden and nonsudden death appear to be alternate outcomes in the same types of patients: (2) only a relatively small number at high risk eventually die within the periods of observation: (3) the high risk patients in the general population differ from the post-myocardial infarction patients. Differences with respect to ventricular arrhythmias themselves include: (1) increased sudden death in patients with ventricular arrhythmia, but the ratio of sudden death to all deaths is not increased: (2) even with low ejection fractions, sudden death requires advanced ventricular arrhythmias: (3) the relative importances of VEB frequency vs. complexity and other critical combinations are not known. (Highgrade ventricular arrhythmia thus is the most obvious practical risk marker, but remains imperfect).

THE VENTRICULAR ECTOPIC BEAT HYPOTHESIS

The relationship of ventricular ectopic beats (VEBs) as the apparent trigger, or at least a marker, for most sudden cardiac death rests on several lines of association: (1) VEBs are related to ventricular tachycardia and ventricular fibrillation: (2) sudden death occurs much more frequently in patients with, than without, VEBs; (3) complex and frequent VEBs are associated with cardiac disease and other risk factors for sudden death: (4) increased frequency and complexity of VEBs are particularly associated with sudden death.

In the Coronary Drug Study[9] there were 256 deaths in three years. Of patients with VEBs of any kind 21.7% died suddenly, while sudden death occurred in only 11.4% of patients without VEBs. The relationship of VEBs to sudden death and to VF in many patients appears to be a function of their frequency and complexity, particularly in the presence of cardiac disease with or without other risk factors. Thus, there is a gross inverse correlation between ejection fraction and recurrent ventricular arrhythmia. Moreover, in coronary disease VRBs are more prevalent and more complex with multi-vessel disease with more proximal obstruction, with increased left ventricular end-diastolic pressure and with asynergy.

Ruberman[10] studied 1,739 male survivors of acute myocardial infarct by one-hour Holter monitoring. Among 202 patients with either runs of VEBs or early cycle VEBs, the five year probability of death from all causes was 37% but for sudden cardiac death was 24.8%. Other complex VEBs were found in 260 patients who had a similar five year probability of death from all causes (33%) but whose sudden death probability was much less, 13.4%. Among 433 patients who only had simple VEBs the sudden death probability was comparable (11.9%), but deaths from all causes was lower, 25.6%. The remaining 244 patients with no VEBs showed comparable (21.4%) death from all causes, but only a 5.6% probability of sudden death. The gradient for all causes primarily reflected the steeper gradient for sudden death. Moss and colleagues[11] also demonstrated the significantly higher risk of complex versus simple VEBs and no VEBs for cardiac death, sudden cardiac death and non-sudden cardiac death.

In 1982 Ruberman[12] compared the five year risk of sudden death in survivors of acute infarction based on the relationship to complex VEBs and early VEBs or runs, to congestive heart failure, to ST segment depression and to heart rate over 90 b/min. The unadjusted risk for complex VEBs and for early VEBs or runs was 2.7 and 3.2 respectively. Adjusted for the other three factors these fell to 2.2 and 2.5. In each VEB group, adjusted or unadjusted risks were greatest with ST depression, next with heart rate over 90 b/min, and finally congestive heart failure. Hypertension was an independent risk factor for sudden death.

VENTRICULAR FIBRILLATION

Ventricular fibrillation (VF), the common mode of sudden death, is considered to be primary when without pump failure or shock; secondary when with failure or shock. Most research has involved patients with ischemic heart disease because this is the most common substrate for sudden death. Of course, other myocardial and valvular diseases are associated with VF[3], often secondary, but sometimes primary. In atrioventricular block and "sick sinus syndrome", most patients are not ischemic, but asystole may terminate in VF.

IDENTIFYING THE RISK OF SUDDEN DEATH

The dominant role of ischemic heart disease produces a risk profile for sudden cardiac death essentially the same as that for coronary
artery disease: age, serum cholesterol, hypertension, obesity and
smoking. With each risk factor, frequent complex and early cycle
(often, but not exclusively R-on-T) VEBs increase the sudden death
risk. Among resuscitated patients, those with acute myocardial infarction (AMI) are at much lower risk than those who are not then
having an infarct. Moreover, the relatively few patients with
hemodynamically important ventricular tachycardia (VT) are at lower
risk than those with VF. Finally, a composite low risk category can
be identified, with about 4% deaths over 10 years - those without
previous myocardial infarction, no significant left ventricular dysfunction, no anterior infarction and no VEBs.

Among patients who have had an AMI, risk of subsequent sudden
death rises with the extent of damage. Yet, VF with a Q wave infarct
(erroneously labelled "transmural") leads to only 2% sudden death in
the first year, while VF with a non-Q wave infarct (erroneously labelled
"subendocardial") produces ten times that risk. Increased risk is
also associated with persistently rapid heart rate, probably related
to reduced function and a hyperadrenergic state.

Following acute myocardial infarction, electrophysiologic studies
indicate that induction of a repetitive ventricular response (RVR) is
related to the risk of sudden death. Greene[13] induced a RVR in 22
patients, of whom 19 developed symptomatic VT of sudden death in one
year. In 39 patients in whom RVR could not be induced, only 4 died
(P>.001).

PROPOSED PREVENTIVE MEASURES

A variety of therapeutic measures appear to reduce the risk of
sudden death. Thus, beta-adrenoceptor blockade has been very encouraging. Anti-arrhythmic treatment without electrophysiological study
- i.e. the vast majority of studies - has been rather dissappointing
and since it is theoretically attractive, this makes questionable
some hypotheses on which the use of primarily antiarrhythmic agents
is based. Other measures include calcium influx blockers to decrease
cardiac automaticity, implanted defibrillators, anti-spastic treatment
(e.g. verapamil, nitrates), surgery (aneurysmectomy; coronary bypass;
endocardial incision and resection), psychologic measures to reduce
aversive stimuli and treatment of risk factors like angina and congestive heart failure.

Lown[1] studied the effect of digitalis in appropriate patients
by giving acetyl strophanthidin, which decreased the frequency and
grade of VEBs and ascribed this to its vagal effect. Lower grades

and infrequent ectopic beats increased, but higher grades were suppressed and in 45% of patients at the time of peak drug action all
ectopic activity was abolished.

Antiarrhythmic therapy must also be considered in the context
of certain metabolic factors, one model of which is myocardial potassium loss. Certain factors promoting both myocardial potassium loss
and arrhythmias are identifiable as related to the risk of dangerous
arrhythmias. These include ischemia, infarction, catecholamine stimulation (especially during hypoxia), hemodynamically important VT, rapid
heart rates, potassium-losing diuretics, and direct current counter-
shock. Thus, one of the actions of some antiarrhythmic drugs like
procainamide, lidocaine and quinidine – decrease of myocardial potassium loss – offers one mechanism for their theoretic effectiveness.

ENDPOINTS OF TREATMENT

Although it is not known whether VEBs are an independent trigger
for VF or merely a marker for more[14] or less[15] important myocardial
abnormality, the endpoints of antiarrhythmic treatment are (1) to
suppress chronic VEBs, especially the more complex and frequent forms;
(2) protect against VT and VF (a) inducible in electrophysiologic
studies, (b) in Holter monitoring and (c) in exercise studies – singly
or combined as needed; (3) to achieve a stable, "adequate", plasma
drug level; (4) to use more than one effective drug if necessary in
patients at higher risk.

PROBLEMS OF ANTIARRHYTHMIC THERAPY

While antiarrhythmic therapy must be aimed at ventricular ectopic
activity (VEA), certain characteristics of VEA and of the antiarrhythmic agents greatly complicate the design and execution of preventive
strategies. In the general population, VEBs are sporadic, rarely symptomatic and very common. Yet, while their frequency/complexity are
related to sudden death, this relation is by no means "clean". Similarly, there is no "clean" association between the mere occurrence of
VEBs and the predisposition to VF. Increased risk is certain for any
group, but many individuals escape and in others, the same grades of
alarming VEA can exist for long periods before developing VF. (What
is the "trigger"?) Choice of the best drug and dose is often difficult although with electrophysiologic study (EPS) it is more likely.
EPS, however, is not practical for the vast majority of people at risk.
Thus, "success" of antiarrhythmic treatment is usually major suppression or abolition of VEA demonstrated by ECG monitoring. However,
such "success" is no guarantee that VF or sudden death will be prevented. Indeed, "spontaneous" variability in VEBs requires very long
monitoring (72-96 hours) for confidence in the treatment, and this
usually is not done. Finally, certain unwanted actions of most

antiarrhythmic drugs can aggravate or induce arrhythmia, and other
side effects often make them unacceptable to the patient.

In considering antiarrhythmic agents to prevent sudden death, it
must be kept in mind that the basis of VF - electrical instability -
is related to a "substrate" of detectable myocardial damage (in most
but by no means all cases)[15]. On this substrate, some precipitant
or "trigger" must act for the patient to die at a particular moment.
The "trigger" may be the VEB, the evidence being not only a statistical association, but the common observation that VEBs, usually early
cycle, lead directly to VF. Thus, suppression of VEBs would prevent
VF if the VEB is the indispensable trigger. On the other hand, if
the VEB is merely a sign of another abnormality, its suppression should
not necessarily prevent ventricular fibrillation. This leads to the
question of whether control of highgrade VEBs prevents sudden death.

Lown[2] studied 72 patients with recurrent VF or VT and hemodynamic
abnormality. Successful treatment of VEA in 46 out of 62 patients
with either coronary or other heart disease resulted in only 3% sudden
death. In 10 patients with no demonstrable heart disease, successful
suppression of VEA completely prevented sudden death during the observation period - N.B. a group with more women and of lower age.
In 16 other patients whose VEA could not be controlled, there was a
37.5% incidence of sudden death - more than 10 times its occurrence
in those successfully treated. Ruskin[16] induced ventricular arrhythmia
in 25 patients in the electrophysiology laboratory. Complete suppression in 19 patients resulted in no deaths during 15 months. In 6
patients in whom ventricular arrhythmia could not be suppressed, three
died suddenly during six months. Van Durme and Bogaert[17] investigated
aprindine for complex VEBs or VT with doses individually adjusted.
Aprindine produced a 60% decrease in complex ventricular arrhythmia
versus a 25% decrease in the placebo group. Yet, sudden death was
not decreased.

Complicating the issue of adding or substituting antiarrhythmic
drug therapy to prevent sudden death is the fact of previous treatment.
Ruskin[16] reported medications at the time of cardiac arrest in 31
patients. Fifteen were taking no cardioactive drugs but 16 had been
taking digoxin, quinidine, quinidine and digoxin and one each procainamide, disopyramide or phenytoin. In his group of 25 patients
with inducible ventricular arrhythmia the range of possible treatments
is suggested by the final agents found to be effective: mexiletene
(9 patients) quinidine (5 patients), amiodarone (3 patients), mexiletene and procainamide (2 patients); quinidine and mexiletene, quinidine and propranolol, quinidine and procainamide, mexiletene and
disopyramide - 1 patient each. In two patients, no antiarrhythmics
were prescribed.

In limited numbers of patients following pre-hospital cardiac
arrest, Myerburg[6,18] showed that achievement of adequate stable

plasma levels of drug was necessary for prevention of VEA and recurrent cardiac arrest. The level of procainamide for preventing VT was
lower in every case than for 85% suppression of VEBs – although these
levels varied considerably among patients, further complicating the
problem. He also studied the risk of recurrent cardiac arrest in a
small population of survivors of pre-hospital cardiac arrest followed
up until a recurrent event, or for at least 24 months. The risk of
recurrent cardiac arrest in those with stable plasma levels of antiarrhythmic agents was 8%, versus 55% in those whose plasma levels fell
into subtherapeutic range or fluctuated markedly. These trends are
encouraging but first need investigation in appropriately designed
prospective trials in adequate numbers of patients.

Another confounding factor for antiarrhythmic therapy is its
variable effect under different physiologic conditions, notably exercise and recovery from exercise. Lown[1] showed tocainide to completely
suppress higher grades of VEBs, despite which prolonged salvos of VT
could occur during and following exercise.

DISSOCIATION BETWEEN VENTRICULAR FIBRILLATION
AND VENTRICULAR ECTOPIC BEATS

The prime problem, perhaps, in considering antiarrhythmic (or
other therapy) of VEBs as our main marker of susceptibility to ventricular fibrillation, is that there is no simple relationship between
VEBs and VF. Thus, antiarrhythmic agents may interfere with any
arrhythmia as a trigger without necessarily abolishing the arrhythmia.
In pre-treated dogs, coronary occlusion produced less VF than in untreated dogs, without a significant decrease in VEBs. Agents like
bretylium and amiodarone suppress VT but not VEBs. In contrast, timolol produced more decrease in sudden death than in arrhythmias.
Moreover, antiarrhythmics may suppress VEA without decreasing mortality. Lidocaine can prevent primary VF, and mexiletene and aprindine can decrease or abolish VEBs, with no effects on mortality, including sudden death.

The results of large scale randomized studies of antiarrhythmic
agents are disturbing. These show that for phenytoin (fixed or variab
dose), tocainide and mexiletene there was a trend in every study for
higher mortality with drug than with control; for aprindine there was
a reverse trend. However, none of these quite reached statistical
significance.

It should be kept in mind the foregoing studies have been large
scale "epidemiologic" trials and that the results of electrophysiologic testing (EPS), previously referred to, remain encouraging.
EPS permits drug screening for safety and effect. It appears that
EPS is best in patients with VT (many with aneurysms or highgrade VEBs
Treatment is carried through to effect of toxicity and can be done
with more than one agent. However, EPS is not practical for widesprea
use.

BETA-ADRENOCEPTOR BLOCKADE AFTER ACUTE MYOCARDIAL INFARCTION

Although beta-blocking agents do not appear to be primarily anti-arrhythmic, compared to large scale antiarrhythmic trials they have had striking success in reducing sudden death and, usually, total mortality. Thus practolol and timolol have reduced both overall mortality and sudden death; propranolol has decreased total mortality and alprenolol has decreased mortality in patients under the age of 65. (The results apply strictly to the doses used)[19].

The attractiveness of postinfarction beta-adrenoceptor blocking trials is due both to reduced mortality and minimal side effects with good patient acceptance. There was little difference between selective agents like practolol and non-selective agents like propranolol, alprenolol and timolol. Moreover, sympathomimetic agents like alprenolol and practolol performed no differently from those lacking this activity like propranolol and timolol. There was also little difference between those with membrane activity like propranolol and alprenolol and those without membrane activity like practolol and timolol - suggesting no direct antiarrhythmic effect. Finally, if we compare antiarrhythmics (excepting aprindine) with beta-blockers, then the weak trend to better results with placebo than with antiarrhythmic intervention is in striking contrast to the strong and significant trend for better results with beta-adrenoceptor blockers versus placebo control.

SUDDEN CARDIAC DEATH: NEEDS

Although treatment of other aspects of cardiac disease may have a distinct impact on sudden cardiac death, experience so far warrants a persistent effort to prevent this catastrophe. Since this necessarily involves widespread, parallel and often cooperative studies, a prime need is the design of protocols, particularly uniform definitions and subsetting of time limits for "sudden death", rhythm categories, functional status and follow-up. Better diagnostic means of predicting VF should be developed. Improved assessment of preventive treatment would include development of antiarrhythmics with decreased side effects, less frequent dosing and, where needed, multiple drug regimens. Parallel efforts should be made to improve myocardial function and ischemia. Further evaluation of neurologic and psychologic contributions to VF susceptibility should also modify these as part of any "trigger" mechanisms. Continued efforts should be made to assess the role of digitalis excess and loss of potassium from the cells. Large scale efforts are needed to evaluate "newer agents" like calcium influx antagonists and bretylium and perhaps other quarternary ammonium agents. Finally, the striking successes of beta-blocking agents raises a strong ethical question of whether these can be omitted in patients at risk. Their concurrent use may be unavoidable and therefore will complicate even more the evaluation of other agents.

TENTATIVE INDICATIONS FOR TREATING VENTRICULAR ARRHYTHMIA

In patients with VEA it is well known that every case need not be treated, particularly if VEBs are infrequent and simple and especially in a context of little or no heart disease. On the other hand, given the present unsatisfactory status of antiarrhythmic therapy, one can develop guidelines for treatment.

Primary VF in the absence of AMI undoubtedly calls for efforts to prevent recurrence, because of the high risk in these patients. Any VEA causing severe symptoms would need to be treated, if only because of the symptoms. Patients with ventricular arrhythmia and coronary disease (angina and/or infarction) who have frequent VEBs and /or grade IV or V VEBs or VT would seem to be prime candidates for effective antiarrhythmic treatment. If exercise or the post-exercise state produces ST depressions of 2mm or more, plus grade IV or V VEBs or VT, treatment for both ischemia and arrhythmia should be instituted. Patients with prolonged QT intervals and VEA, particularly those who have had syncope, are in great danger of malignant ventricular arrhythmias and should be treated. Finally, in mitral (or tricuspid) prolapse and in obstructive cardiomyopathy with syncope and grade IV or V VEBs or VT it would seem prudent to attack the ventricular ectopic activity.

REFERENCES

1. B. Lown, Sudden cardiac death: The major challenge confronting contemporary cardiology. Am. J. Cardiol. 43:313-328, (1979).
2. B. Lown and R. B. Graboys, Management of patients with malignant ventricular arrhythmias. Am. J. Cardiol. 39:910-918 (1977).
3. D. H. Spodick, Effective management of congestive cardiomyopathy: Relation to ventricular structure and function. Arch. Inter. Med. 142:689-692 (1982).
4. S. Goldstein, R. Landis, R. Leighton, G. Ritter, C. M. Vasu, A. Lantis and R. Serokman, Characteristics of the resuscitated out-of-hospital cardiac arrest victim with coronary heart disease. Circulation 64:977-984 (1981).
5. L. E. Hinkle and H. T. Thaler, Clinical classification of cardiac death. Circulation 65:457-464 (1982).
6. R. J. Myerburg, C. A. Conde, S. J. Sung, A. Mayorga-Cortes, S. M. Mallon, D. S. Sheps, R. A. Appel and A. Castellanos, Clinical, electrophysiologic and hemodynamic profile of patients resuscitated from prehospital cardiac arrest. Am. J. Med. 68:568-576 (1980).
7. A. A. J. Adgey, J. E. Devlin, S. W. Webb and H. C. Mulholland, Initiation of ventricular fibrillation outside hospital in patients with acute ischaemic heart disease. Br. Heart J. 47: 55-61 (1982).

8.   R. J. Myerburg, L. Zaman, K. M. Kessler and A. Castellanos,
     Evolving concepts of management of stable and potentially
     lethal arrhythmias. Am. Heart J. 103:615-625 (1982).
9.   S. Goldstein, Necessity of a uniform definition of sudden coronary
     death: Witnessed death within 1 hour of the onset of acute
     symptoms. Am. Heart J. 103:156-159 (1982).
10. W. Ruberman, E. Weinblat, C. W. Frank, J. S. Goldbert and S.
     Shapiro, Repeated 1 hour electrocardiographic monitoring of
     survivors of myocardial infarction at 6 month intervals:
     Arrhythmia detection and relation to prognosis. Am. J. Cardiol.
     47:1197-1204 (1981).
11. A. J. Moss, H. T. Davis, J. DeCamilla and L. W. Bayer, Ventricular
     ectopic beats and their relation to sudden and nonsudden cardiac
     death after myocardial infarction. Circulation 60:998-1003
     (1979).
12. W. Ruberman, E. Weinblatt, J. D. Goldbert and S. W. Frank, Sudden
     death after myocardial infarction: Runs of ventricular prema-
     ture beats and R-on-T as high risk factors. Am. J. Cardiol.
     45:444 (1980).
13. H. L. Greene, P. R. Reid and A. H. Schaeffer: The repetitive ven-
     tricular response in man. New Engl. J. Med. 299:729-734 (1978).
14. G. J. Taylor, J. O. O'Neal, B. Pitt, S. C. Griffith and S. C.
     Achuff, Complex ventricular arrhythmias after myocardial in-
     farction during convalescence and follow-up: A Harbinger of
     multi-vessel coronary disease, left ventricular dysfunction
     and sudden death. John Hopkins Med. J. 149:1-5 (1981).
15. T. Sugiura, A. P. Flessas, J. S. Singh and D. H. Spodick, Ventric-
     ular arrhythmias in the late hospital phase of acute myocardial
     infarction. Circulation 64(IV) IV-306 (1981).
16. J. N. Ruskin, J. P. DiMarco and H. Garan, Out-of-hospital cardiac
     arrest. Electrophysiologic observations and selection of long-
     term antiarrhythmic therapy. New Engl. J. Med. 303:607-613
     (1980).
17. J. P. VanDurme, M. G. Bogaert, Prevention of sudden death: Role
     of antiarrhythmic therapy. In: H. E. Kulbertus and H. J. J.
     Wellens, Sudden Death, DenJuag, Nijhoff, pp324-327 (1980).
18. R. J. Myerburg, F. W. Briese, C. Conde, S. M. Mallon, R. R. Liber-
     thson and A. Castellanos: Long-term antiarrhythmic therapy
     in survivors of prehospital cardiac arrest. Initial 18 months
     experience. JAMA 238:2621-2624 (1977).
19. J. R. Hampton and M. A. D. Phil, Beta-blockade and the secondary
     prevention of myocardial infarction. Acta Med. Scand. Suppl.
     651: 219-226 (1981).

POSSIBILITIES OF PHARMACOLOGICAL PREVENTION OF SUDDEN DEATH

Lars Wilhelmsen

Department of Medicine
Östra Hospital
S-416 85 Göteborg
Sweden

INTRODUCTION

Coronary heart disease (CHD) is the dominating cause of death
in most industrialized countries. Sudden coronary death (SD) has
been found to be the most common type of death both in the early
phases of CHD and during several years' follow-up. There is reason
to believe that most of these sudden deaths are due to ventricular
tachyarrhythmias.

It is usually found that the same risk factors are present in
persons subject to non-fatal and fatal CHD, and more specifically
to SD. Thus, the risk of SD has been related to smoking habits,
hypertension, hypercholesterolemia and in some studies also to
obesity, diabetes, lack of physical activity and ECG changes. Some
findings indicate more profound aberrations in those who die from
CHD as compared to those who survive the initial attack (Tibblin
et al., 1975). Alcoholic intemperance has also been associated with
SD (Wilhelmsen et al., 1973, Elmfeldt et al., 1976).

Ventricular premature beats at rest (Kannel and Gordon 1973)
or during exercise (Chiang et al., 1969) do not seem to be associated with any increased risk of SD in otherwise healthy individuals,
but were associated with increased risk in subjects with a diseased
myocardium as evidenced by associated ECG abnormalities (Vedin et
al., 1972, Kannel and Gordon 1973).

In addition to number and size of previous infarctions a number
of studies have shown ventricular ectopic beats to be of significance
for the outcome after an MI. It should be stressed that the arrhythmias have been recorded after discharge from hospital, or just

131

prior to discharge. Arrhythmias detected during the acute coronary
care unit period have not been of significant importance. Cats et
al., (1979) have stressed that ventricular premature complexes are
so common in post-infarction patients that the clinical usefulness
of this finding is questionable. It is probable that we need a
better index as to which type of premature beats carry prognostic
information. This information might only be found during some type
of electrophysiological measurement, perhaps including some type of
ischemic provocation.

When studying a general population sample of middle-aged men
(Wilhelmsen et al., 1972) we also found that 40% of SD victims had
suffered clinically verified MI or angina pectoris before death.

In studies at the special Post-MI Clinic in Göteborg, Sweden,
which started in 1968 (Elmfeldt et al., 1975), the principle has
always been to study non-selected infarction patients from hospital
treatment throughout the entire disease until death. In addition,
all out-of-hospital events have been followed with the aid of a
special Ischaemic Heart Disease Register (Elmfeldt et. al., 1975).
The total autopsy rate for patients up to 65 years who have been
studied has been 90% or higher.

Among infarct patients it was found that about two-thirds of
deaths up to two years after the infarction occurred within 24 hours
of onset of symptoms. About one-half occurred outside the hospital.
According to autopsy, two-thirds of the deaths were caused by a
fresh infarction. One quarter of the deaths were ascribed to malignant arrhythmias. In these cases, autopsy did not reveal any fresh
infarctions (Vedin et al., 1975).

Possibilities of Pharmacological Prevention

When discussing various possibilities of prevention we have to
consider both long-term and short-term interventions. Disturbances
of the vessel wall which can be atherosclerosis or coronary artery
spasm are certainly responsible for the background susceptibility
to arrhythmias in most CHD patients. Thrombotic mechanisms probably
play a role in some cases. However, myocardial factors including
disturbed metabolism and changes of the membrane potentials of the
myocardial cells are important. Such changes are supposed to be
the immediate initiators of various cardiac arrhythmias responsible
for SD. It is also accepted that most of these deaths are due to
cardiac tachyarrhythmias. Since SD occurs very shortly after the
onset of any type of symptoms, therapeutic agents should, ideally
already be in use when the acute process starts in order to prevent
a fatal outcome.

The potential benefits of drugs affecting blood lipids, the
coagulation system, or the thrombocytes have attracted great interest

for some time, but the present balance of evidence does not seem to favor their use in prevention of SD in the short-term perspective. In the long-term perspective they might, however, be important as well as primary preventive measures. The balance between possible benefits and risk of side effects has to be considered in any form of preventive treatment. In post-MI patients at high risk of death, drug treatment including risk of adverse effects may well be accepted, but such treatment may be contraindicated in healthy individuals even if arrhythmias are present.

Antiarrhythmic drugs

Other drugs than beta-blockers with antiarrhythmic effects, such as quinidine, procainamide, and phenytoin, as well as other, newer agents, have not been subjected to comprehensive, long-term studies of the effect on SD. Kosowsky et al., (1973) found only a limited reduction in ventricular premature beats with use of quinidine or procainamide, and a high frequency of side effects. However, Jones et. al., (1974) found a significant reduction in ventricular premature beats after MI during treatment with quinidine. Phenytoin was tested during one year after MI by Lowell et. al., (1971) who found a slight, insignificant effect on ventricular premature beat frequency and no effect on mortality in 283 patients randomly allocated to the active drug, compared with 285 patients given placebo. Peter et. al., (1978) studied 150 post-MI patients and did not either find an effect of phenytoin.

Several of the older drugs with antiarrhythmic effects (procainamide, quinidine, amiodarone) are not suitable for long-term use, at least not in patients with low - moderate risk of SD because of side effects.

Disopyramide, mexiletine, tocainide, and aprindine are the main drugs presently being investigated for prolonged therapy of serious ventricular arrhythmias (Campbell et al., 1977, 1978; Engler et al., 1979; Hagemeijer et al., 1979; Koch-Weser, 1979; Winkle et al., 1978). They have been shown effective in preventing arrhythmias. Although side effects leading to discontinuation of therapy have been relatively common, the severity has not been to such a degree that they are considered unsuitable in high risk postinfarction patients.

Aprindine has been studied in a randomized trial in 305 survivors of recent MI (Hagemeijer et. al., 1979). In many patients serious ventricular arrhythmias disappeared during therapy, but data on SD and documented ventricular fibrillation were disappointing; 12 SD:s in the aprindine group (n=153), and 19 SD:s in the placebo group (n=152) according to the most recent information (van Durme, personal commun., 1982). Drug intolerance caused a decrease of the

dose level in several cases, and it is possible that this dose decrease had to be done to a level where the drug was no longer effective, or its mode of action was no longer appropriate.

Tocainide was studied in 162 post-MI patients by Rydén et al.,
(1980), and by Bastian et al., (1980) in 146 post-MI patients but
without any effect on mortality. Similarly, Chamberlain et al.,
(1980) were unable to find a protective effect on 344 patients
randomized to mexiletine or placebo.

In summary, the purely antiarrhythmic drugs do not at present
seem to offer great hope for the majority of post-MI patients, but
might be of value in selected cases.

Concluding remarks

SD often hits subjects who have previously suffered clinical
coronary heart disease, especially MI. In this latter group of
patients there are great differences in prognosis, and among those
at high risk there is a great need for secondary preventive measures.
Since most deaths occur soon after the onset of symptoms, therapeutic
agents should ideally already be present in the body in sufficient
concentrations when the process starts.

The potential benefits of drugs affecting blood lipids, the
coagulation system, or the thrombocytes have attracted great interest
but for short-term prevention of SD they have so far not been shown
effective. Purely anti-arrhythmic drugs may be of value in selected
cases, but in general their side-effect pattern does not encourage
more general use in patients with low-moderate risk of SD. Thus,
there is so far no conclusive evidence of a secondary preventive
effect of long-term treatment with anti-arrhythmic drugs other than
beta-blockers.

REFERENCES

Bastian, B.C., McFarland, P. W., McLauchlan, J. H., Ballantyne, D.,
        Clark, R., Hillis, W. S., Rae, A. P., Hutton, I., 1980, A
        prospective randomized trial of tocainide in patients
        following myocardial infarction. Am. Heart J., 100:1017.
Campbell, R.W.F., Talbot, R. G., Julian, D. G., Prescott, L. F.,
        1977, Long-term treatment of ventricular arrhythmias with
        oral mexiletine. Postgrad. Med. J. 53(1):146.
Campbell, N. P. S., Pantridge, J. F., Adgey, A. A. J., 1978, Long-
        term oral anti-arrhythmic therapy with mexiletine. Br. Heart
        J. 40:796.   ·
Cats, V. M., Lie, K. I., van Capelle, F. J. L., et. al., 1979,
        Limitations of 24 hour ambulatory electrocardiographic re-
        cording in predicting coronary events after acute myocardial
        infarction. Am. J. Card. 44:1257.

Chamberlain, D. A., Jewitt, D. E., Julian, D. G., Campbell, R. W. F.,
    Boyle, D. McC. , Shanks, R. G. 1980, Oral mexiletine in high-
    risk patients after myocardial infarction. Lancet 2:1324.
Elmfeldt, D., Wilhelmsen, L., Tibblin, G., Vedin, J. A., Wilhelmsson,
    C., Bentsson, C. 1975, A postmyocardial infarction clinic in
    Göteborg, Sweden. Acta. Med. Scand. 197:497.
Elmfeldt, D., Wilhelmsen, L., Tibblin, G., Vedin, J. A., Wilhelmsson,
    C., Bengtsson, C. 1975, Registration of myocardial infarction
    in the city of Göteborg, Sweden. J. Chron. Dis. 28:173.
Engler, R., Ryan, W., LeWinter, M., Bluestein, H., Karliner, J. S.,
    1979, Assessment of long-term antiarrhythmic therapy: Studies
    on the long-term efficacy and toxicity of tocainide. Am. J.
    Cardiol. 43:612.
Hagemeijer, F., Van Durme, J. P., Lubsen, J., et. al., 1979, The
    Ghent-Rotterdam aprinidine study; antiarrhythmic prophylaxis
    after myocardial infarction. In: Florence International
    Meeting on Myocardial Infarction, Excerpta Medica.
Jones, D. T., Kostuk, W. J., Gunton, R. W. 1974, Prophylactic quin-
    idine for the prevention of arrhythmias after acute myocardial
    infarction. Am. J. Cardiol. 33:655.
Kannel, W. B., Gordon, T. 1973, Assessment of coronary vulnerability
    - The Framingham Study. In Early Phases of Coronary Heart
    Disease. The Possibility of Prediction. J. Waldenström,
    T. Larsson & N. Ljungstedt, eds., Skandia International Sym-
    posia, Stockholm. p. 123.
Koch-Weser, J. 1979, Disopyramide. N. Eng. J. Med.300:957.
Kosowsky, B. D., Taylor, J., Lown, B., Ritchie, R. F. 1973, Long
    -term use of procaine amide following acute myocardial in-
    farction. Circulation 47:1204.
Lowell, R. R. H., et. al., 1971, Phenytoin after recovery from myo-
    cardial infarction. Lancet II:1055.
Peter, T., Ross, D., Duffield, A., Luxton, M., Harper, R., Hunt, D.,
    Sloman, G. 1978, Effect on survival after myocardial infarction
    of long-term treatment with phenytoin. Br. Heart J. 40:1356.
Ryden, L., Arnman, K., Conradson, T-B., Hofvendahl, S., Mortensen, O.,
    Smedsgard, P. 1980, Prophylaxis of ventricular tachyar-
    rhythmias with intravenous and oral tocainide in patients
    with and recovering from acute myocardial infarction. Am.
    Heart J. 100:1006.
Tibblin, G., Wilhelmsen, L., Werko. 1975, Risk factors for myocardial
    infarction and death due to ischemic heart disease and other
    causes. Am. J. Cardiol. 35:514.
Vedin, A., Wilhelmsson, C., Wilhelmsen, L., Bjure, J., Ekström-
    Jodal, B. 1972, Relation of resting and exercise-induced
    ectopic beats to other ischemic manifestations and to coronary
    risk factors. Am. J. Cardiol. 30:25.
Vedin, A., Wilhelmsson, C., Elmfeldt, D., Säve-Söderberg, J.,
    Tibblin, G., Wilhelmsen, L. 1975, Deaths and non-fatal re-
    infarctions during two years' follow-up after myocardial
    infarction. Acta. Med. Scand. 198:353.

Wilhelmsen, L., Tibblin, G., Werko, 1972, A primary preventive study
        in Gothenburg, Sweden. Preventive Medicine 1:153.
Winkle, R. A., Meffin, P. J., Harrison, D. C. 1978, Long-term
        tocainide therapy for ventricular arrhythmias. Circulation
        57:1008.

MYOCARDIAL INFARCTION AND SUDDEN DEATH AFTER CORONARY ARTERY SURGERY

Sergio Dalla-Volta, Federico Corbara, and Brenno Permutti

Dept. of Cardiology
of the University of Padova

Vincenzo Gallucci and Dino Casarotto

Dept. of Cardiac Surgery
of the University of Padova

The autologous saphenous vein bypass graft (CABG) was introduced in 1969 with the aim of adding a method for treatment of anginal pain and with the hope of reducing the long term appearance of myocardial infarction (MI) or reinfarction and sudden coronary death (SD).

The physiological basis of these goals were founded on the possibility of correcting the consequences of the critical reduction of the cross section area of the coronary vessels, through the restoration of an adequate modular blood flow to the ischemic myocardium.

Twelve years of experience with surgery have not completely settled the controversies between the advocates of medical or surgical treatment, as the effective possibility of the CABG of preventing at large MI and SD is not proved. The consensus is more uniform on the action of surgery for the relief of angina pectoris, as surgery in most series gives better results, at five and ten year follow-ups in comparison to medical treatment.

These conflicting opinions partly rely on the scarcity of data analyzing the long term possibilities of the prevention of a late MI and SD in a large population of patients with ischemic heart disease operated with CABG. It seemed therefore desirable to review the fate of patients who had survived coronary artery surgery long enough to permit a critical reassessment of the topic.

Material and Method

For this study two different groups of patients were reviewed, totalling 734 patients who had survived to the CABG operation performed for relief of angina pectoris (at effort or spontaneous) associated with a critical (more than 75%) narrowing of one or more coronary vessels. Excluded from this analysis were operated patients with left ventricular aneurysm or other cardiac lesions, usually aortic valve disease, combined to the stenosis of one or more coronary vessel (s).

a)   The first group comprised 211 consecutive patients (191 males and 20 females) operated in the Dept. of Cardiac Surgery of the University of Padova between 1974 and 1977: these patients have subsequently been followed in The Cardiology Dept. for at least five years.

b)   The second group is represented by 523 consecutive patients (496 males and 27 females) or age ranging from 30 to 75 years (mean 51.2 years) operated in the years 1975 and 1976 in the Dept. of Cardiac Surgery of the University of Padova at Padova and Verona and the Hospital "Niguarda" of Milan: these patients have been reassessed five years after the operation.

In the two groups of patients several parameters have been analyzed and the overall results published elsewhere[1]. For the goals of this study the following parameters, both before and after operation, have been reviewed: age and sex; severity of the lesion, calculated and number of coronary vessels involved by the disease; subjective symptoms; conditions of the left ventricle at left ventricular angiography; degree of revascularisation obtained by the operation. The incidence of MI has been separately evaluated before the CABG, during the operation and the first month, and after this period. Variations of anginal pain have been carefully evaluated. The incidence and the causes of death, the latter divided in non cardiac, cardiac and sudden, assessed with the aid of a questionnaire sent to the families and to the physician of the deceased patients. The type and frequence of arrhythmias were monitored with dynamic ECG for twenty four hours; the reassumption or not of a regular working activity and the response to effort matched with the clinical conditions of the patients (included the level of sexual activity described in a separate paper).

From this analysis the patients deceased in the immediate postoperative period were excluded (hospital mortality in the first thirty days): the incidence of these deaths is 4.5%.

The survival curves have been calculated with the method suggested by Cutler and Ederer[2].

These data have been matched with the clinical picture, the patterns of the left ventricle and coronary vessels before the operation.

All surviving patients were controlled periodically with clinical, electrocardiographic and chest X-rays studies.

Results

Angina: five years after CABG 64.5% of patients from the first group and 77.9% from the second group were asymptomatic, while only 6.6% in each group complained that the angina was "similar or worse," than before the operation. The reappearance or the worsening of the cardiac pain occurred in 51% of the subjects in the first six months after surgery, in the remaining 49% it was distributed through the total observation period (Table 1).

No correlation was observed between the changes of the pain and the different preoperative parameters.

Arrhythmias: the twenty fours dynamic electrocardiogram demonstrated in 25% of different types of arrhythmias, while 75% of patients were free.

In the arrhythmic group, 24.7% of patients had sporadic supraventricular premature beats; 2.4 had ventricular ectopic beats more frequent than one every fifteen, but less than ten beats. In 4.8% of cases the arrhythmias were represented by atrial parosysmal fibrillation, in 1.1% ventricular tachicardia with narrow QRS complexes. Most patients (68%) were unaware of the presence or absence of rhythm troubles.

Comparison of the behavior of angina with the presence or not of arrhythmias shows that in 42% of cases with recurrence of pain at least one type of arrhythmia was detected at Holter monitoring, while 20.6% of patients had arrhythmias and no pain.

No correlation was observed between a single type of arrhythmias and the presence or absence of angina pectoris.

Not fatal, late myocardial infarction: six cases in the first group (3%) and eleven in the second (2%) had documented not fatal myocardial infarction. In eight of the seventeen patients (47%) MI was observed in subjects who had a cardiac infarction before surgery. The annual incidence of MI in the post-operative history of bypassed patients was less than 0.5%. The figures reach 1.1% if the cases deceased from cardiac death in whom the possibility of an episode of MI as cause of death is likely are included.

No correlation was observed between any of the pre-operative parameters and the incidence of late MI.

Late death: eleven patients of the first group (5.2%) and twenty-eight of the second group (5.3) deceased late after operation

SD, in the long term follow-up. As the mortality rate in the years
1974-1975 was around 5% a certain number of "would be" cases of SD
are subtracted from the statistics. Only a randomized study, with
patients at the same risk listed in different series, will answer the
question.

Indirectly some suggestions can be cautiously raised. Cardiac
ischemia is the leading cause of electrical instability, the basic
mechanism of the ventricular fibrillation and surgery is effective in
the reduction of the other clinical manifestations of cardiac
ischemia. The disappearance or the reduction of angina, the improvement of the ergometric test and of cardiac perfusion, as assessed by
isotopic techniques, and the linear correlation between the clinical
results and the completeness of cardiac revascularization after
surgery are indirect evidence of the improvement of ischemia.

It seems therefore conceivable that the CABG has the potential
of reducing the rate of ventricular fibrillations and of cardiac SD.
The small number of cardiac and SD in this study are associated with
the observation that 50% of cardiac death (sudden and not) occurred
in the first year after surgery. This is the critical period of
coronary artery surgery, during which the highest numbers of closure
of grafts and of SD are numbered.

After bypass the functional improvement of the left ventricle is
partial and absent in presence of a large area of necrotic tissue.
These regions, akynesiea and dyskinesia are the trigger areas if the
electrical instability of the heart, justifying the reason why the
most important correlate of cardiac death has been in one series the
preoperative function of the left ventricle, as expressed by the
ejection fraction.

In spite of several unsolved problems, from this study two
positive conclusions can be drawn: besides the significant improvement of the symptoms, the most deadful cardiac events (MI and cardiac
death, sudden and not) are rather infrequent and the incidence is
quite comparable to the total figures of the Italian population of
the same age, in the same period.

REFERENCES

1. S. Dalla-Volta, F. Corbara, P. Stritoni et al., "Risultati a
       Distanza del bypass aortocoronarico" G.Ital.Cardiol.in Press.
2. S. J. Cutler, F. Ederer "Maximum utilization of the life table
       method in analyzing survival" J.Chron.Dis. 8:699, (1958).
3. L. Campeau, J. Lésperance, J. Hermann, F. Corbara, C. M. Grondin,
       M. G. Bourassa, "Loss of the improvement of angina between 1
       and 7 years after aortocoronary bypass surgery. Correlations
       with changes in vein grafts and in coronary arteries." Circu-
       lation 60: I-1, (1979).

4. L. A. Vismara, R. R. Miller, J. E. Price, R. Karem, A. N. Maria,
    D. T. Mason, "Improved longevity due to reduction of sudden
    death by aortocoronary bypass in coronary atherosclerosis.
    Prospective evaluation of medical versus surgical therapy in
    matched patients with multivessels disease." Am.J.Cardiol 919:
    39, (1977).
5. M. G. Bourassa, P. E. Puddu, J. Helias, N. Danchin, G. Goulet, P.
    David, "Possibilité de prevention de la mort subite par le
    pontage aorto-coronarien." Ann.Cardiol.Angéiol 29:237, (1980).
6. K. E. Hammerneister, T. A. Rouen, J. A. Murray, H. T. Dodge.
    "Effect of aortocoronary saphenous vein bypass grafting on
    death and sudden death." Am.J.Cardiol. 39:925, (1977).

ASYMPTOMATIC MYOCARDIAL ISCHEMIA IN CORONARY

PATIENTS WITH STABLE ANGINA PECTORIS

Giuseppe Cocco*, Carlo Strozzi,
Raffaele Pansini, Barbara Leishman, and
Claudio Sfrisi

*Department of Internal Medicine
Kantonsspital Basel, Switzerland, and
Cardiological Department
Medical and Therapeutical Clinics
University of Ferrara, Arcispedale S. Anna
I-44100 Ferrara, Italy

SUMMARY

Little information is available on the frequency and characteristics of transient myocardial ischemia (TMI) in patients with
stable angina pectoris. We selected 40 coronary patients with stable
and typical angina. The presence of coronary artery disease and the
ejection fraction were evaluated by means of angiocardiography.
Dynamic electrocardiographic monitoring (DCG) was performed with
bichannel portable recorders for three 24 hour periods at 7-10-day
intervals. The patients were on optimal and stable doses of β-
blockers and isosorbidilate throughout the whole study. The DCG
analyzer and the method of analysis were especially adapted for
this study (S-T segment analysis).

We detected 788 episodes of TMI in 22 of the 40 patients. The
ejection fraction was poorer in the 22 patients with ST-T changes than
in the 18 without. Furthermore, the incidence of angina pectoris was
higher in the 22 patients with ST-T changes than in the 18 without.

The repolarization changes were:

(1) ST elevation (55 symptomatic and 87 asymptomatic episodes);
(2) ST depression (138 symptomatic and 236 asymptomatic episodes);
(3) T-wave changes (83 symptomatic and 164 asymptomatic episodes).

All 22 patients with TMI presented a combination of the above changes.

It appears, therefore, that the ST-T changes are more frequent in patients with stable and typical angina pectoris than was hitherto suspected. The DCG is valuable in assessing these changes, especially when one considers that the asymptomatic episodes are almost twice as frequent as the symptomatic ones.

The asymptomatic episodes lasted for a mean of 1.8±1.3 min (mean±S.D.), while the symptomatic episodes lasted 3.8±2.7 min (p<0.002 by sign test). Heart rate was unchanged during the episodes of TMI and did not show any significant difference between asymptomatic and symptomatic episodes. Additional investigation is however necessary to determine the clinical implication of these findings.

INTRODUCTION

In recent years, several investigators[1-10] have demonstrated that the serial dynamic ECG (DCG) can indentify episodes of transient myocardial ischemia (TMI) in a large percentage of patients with coronary artery disease (CAD). Furthermore, it has been shown that only a minority of these ischemic episodes are accompanied by chest pain or equivalent symptoms[2,4-10]. It is therefore possible that the usual criterion of chest pain and effort-induced ECG changes may underestimate the actual frequency of TMI[1-4].

Most studies, however, have been in patients with unstable[3,5,7,9,10,14] or variant angina pectoris[6,11,12,13]. Only a few studies have been performed in patients with stable angina pectoris[1,2,4] and these have been relatively simple in design. Furthermore, in some of these studies the authors used 1-channel DCG recorders and analyzed the tracings with first generation decoders, a situation far from ideal for the detection of TMI. The aim of the present investigation was therefore to increase our knowledge about the frequency and characteristics of episodes of TMI in patients with CAD and stable angina pectoris.

MATERIAL AND METHOD

We studied 26 male and 14 female patients with stable angina pectoris and CAD who were referred to our department for routine check-ups. Their age was 59.6±7.1 years (mean±S.D.) and the diagnosis of CAD had been known for a mean of 3.9 years (range 1.3-6.0). Their angina pectoris was stable (unchanged characteristics for a mean period of 6.2 months, range 4.5-8.3 months) and typical (ie. effort-induced and reproducible). Patients with arterial hypertension were excluded because we were unable to check their 24-hour blood pressure (BP) under ambulatory conditions. Patients with diabetes mellitus were also excluded, because we ignored the possible effect of the disease on the incidence and type of TMI. Patients with relevant valvular diseases were also excluded. At the time

of this study all patients were on "optimal" doses of β-blockers
(metoprolol in 31 cases, oxprenolol in 9 cases) and isosorbidilate.
Optimal dose is defined as that dose which reduces the frequency
and severity of effort-induced and spontaneous angina pectoris without
inducing important side effects. The dosage of antianginal drugs
had been stable for a mean of 3.9 months (range 3.0-4.2 months) prior
to the study. The antianginal drugs were administered by an independent group of physicians. Selective coronary angiography was performed 4 to 15 months (mean 5.1) before this study. The coronary
angiograms were considered to be pathological in all patients: 10
patients presented more than 60% narrowing of one major coronary
artery, 21 patients presented two-vessel CAD and 9 patients presented
three-vessel CAD. None of the patients had left main CAD or obstuc-
tion of two vessels including the proximal descending coronary artery.
The left ventricular function was evaluated angiographically from a
right anterior oblique view of 45° and was classified as normal
(ejection fraction 60% or more), reduced (40-59%) or poor (39% or
less).   26 patients, including one patient with right bundle branch
block and three patients with first degree atrioventricular block,
had a normal (ie. non ischemic) resting ECG. 11 patients had signs
of old myocardial infarction (diaphragmatic in 7, anterior in 4)
and three patients presented ST depression (0.5mm in V_{5-6}). Ambulatory ECG monitoring (DCG monitoring) was performed using 2-channel
AdvanceMed recorders, model 2600. The tapes were analyzed by means
of the scanner AdvanceMed model 7000 eliminator. However, the original AdvanceMed transcriptor was replaced by a Delalande transcriptor,
model Dyna D 800, to avoid distortion of the ECG[15,16]. DCG monitoring
was performed several times during 24 hours of normal daily activities. The electrodes were applied by means of the Quinton Gun.
Three 24 hr DCG monitorings per patient were obtained at 7-10-day
intervals. Each patient was instructed to keep a detailed diary of
activities and symptoms. The beginning and the end of symptoms were
to be marked on the tapes by pushing an event marker. We used the
derivations suggested my MacAlpin, ie. CC_1 and CC_3[11,12], in 10
patients, because in a previous running period we observed that
these 10 patients presented more ST-T changes in V_3 than in V_5.
In the other 30 patients, however, we used the derivations shown
in Figure 1, because we have found these derivations to be the best
for the study of ST-T changes[16]. The tapes were scanned visually
at 240 times real time speed and an integral 24 hr trend transcrip-
tion was obtained. The ischemic episodes were counted and classified. An ischemic episode was regarded to be present when:

(1) ST depression with horizontal or downsloping morphology of at
    least 1.5mm from the resting level and of at least 80msec dur-
    ation;
(2) a slow upsloping ST segment depressed at least 2.0mm 80msec after
    the J point;
(3) a slow upsloping or horizontal ST segment elevation of at least
    1.5mm and 80msec duration;

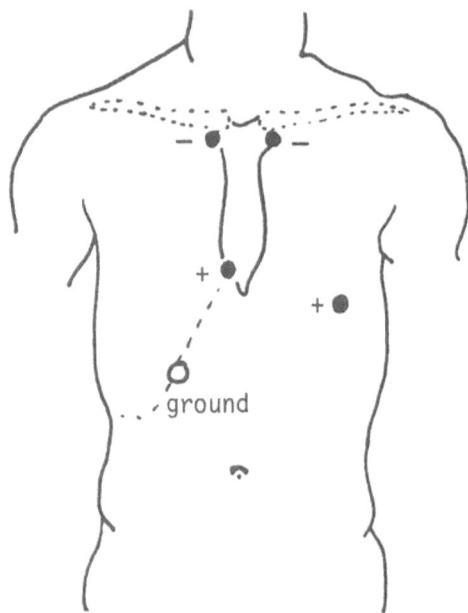

Fig. 1.   Schematic representation of lead placement used. For
          channel 1, the negative electrode is placed to the right
          of the sternal manubrium and the positive electrode is
          placed immediately to the right of the sternum on the
          4th-5th intercostal space. The ECG resembles V_1 or V_2.
          For channel 2, the negative electrode is placed to the
          left of the sternal menubrium and the positive electrode
          in the V_5 or V_6 position. The resulting QRST complex re-
          sembles V_5 or V_6. The first electrode (ground) is placed
          as shown in the figure or, in rare cases, in the V_{5R}
          position.

(4) inversion or pseudo-normalization of the T-wave,

these changes were observed in six or more consecutive ECG complexes.
These features are widely accepted as indicators of myocardial ischemia[17-19]. For the purpose of this study, the DCG monitoring
was not regarded as positive if only transient rhythm disturbances
were detected, in spite of the fact that some of these arrhythmias
may indeed be associated with myocardial ischemia.

RESULTS

Negative DCG Monitoring

     In 18 of the 40 patients, the DCG monitoring did not detect
any ischemic episodes. In these patients, the ejection fraction was

normal in 9, reduced in 6 and poor in 3 cases. 15 of the 18 patients
had effort-induced angina pectoris on other days when DCG monitoring
was not performed. On the other hand, with the treatment the frequency of angina pectoris was low: a mean of 1.2 episodes per week
(range 0.5-2.5 episodes/week), and spontaneous angina pectoris was
exceptional.

Positive DCG Monitoring

     788 episodes of TMI were detected in 22 patients. ECG changes
of the following types were observed:

(1) ST elevation: 55 symptomatic and 87 asymptomatic episodes;
(2) ST depression: 138 symptomatic and 236 asymptomatic episodes;
(3) T-wave changes (in 66% of cases pseudo-normalization): 83 symp-
    tomatic and 164 asymptomatic episodes.

In 17/22 patients, a combination of the above ST changes were associ-
ated with T-wave changes, while in 5/22 patients two or more of the
ST-T changes alternated. When associated, T-wave changes preceded
ST changes. Thus chest pain or equivalent symptoms (thoracic op-
pression or dyspnea) were reported to occur in 276/788 episodes of
ST-T changes (35%), while 512 episodes (65%) were asymptomatic.

     These 22 patients with positive DCG monitoring reported a mean
of 5.2 episodes/week of angina pectoris (range 2.5-6.5), clearly
more than the 18 patients with a negative DCG monitoring. The ejection fraction in these 22 patients was normal in 6, reduced in 4
and poor in 12 cases.

     The asymptomatic episodes were almost twice as frequent as the
symptomatic ones. The ST-T changes were similar in both symptomatic
and asymptomatic episodes. Symptomatic and asymptomatic episodes
appeared to be present in all 22 patients. In symptomatic episodes,
chest pain generally followed a mean of 2.9min (range 20sec - 4min)
after the appearance of the ECG changes. In 14 episodes, the pain
appeared (as shown by the event marker on the tapes) 2-5min after
the ST-T changes disappeared. In 18 cases the patients noticed ischemic symptoms, but no ST-T changes were detected 5min prior or
after the event marker. The symptomatic episodes lasted for
3.8±2.7min (mean±S.D.) and the asymptomatic episodes for 1.8±1.3min,
the difference being statistically highly significant (p<0.002, by
sign test). During the ST-T changes the heart rate (HR) was usually
normal and showed no relevant differences between symptomatic
69.6±14.8 beats/min) and asymptomatic episodes (62.5±18.3). Tachycardia (HR>100) was observed in association with 11 ST-T changes
(4 symptomatic and 7 asymptomatic). All data are presented in
Table 1.

Table 1. ST-T Changes in 22 Patients

Transient Myocardial Ischemia	Symptomatic	Asymptomatic	Total
No. of ST-T changes	276 (35%)	512 (65%)	788 (100%)
Duration of changes (min.)*	3.8 ± 2.7**	1.8 ± 1.3**	2.4 ± 2.9
Types			
T-wave changes	83 (10.5%)	169 (21.4%)	252 (32%)
ST elevation	55 (7.0%)	87 (11.0%)	142 (18%)
ST depression	138 (17.5%)	256 (32.5%)	394 (50%)
Heart rate (bpm)*	69.6 ± 14.8	62.5 ± 18.3	66.1 ± 13.4

  * Mean ± 1 S.D
** p<0.002, Sign test

Figure 2 shows an example of asymptomatic TMI and figure 3 an
example of symptomatic TMI.

DISCUSSION

Our data show that some patients with stable angina pectoris and
CAD present ST-T changes that are similar to those described in
patients with variant[6,11,13] or unstable angina pectoris[3,5,7,9,10,14].
These ECG changes are detectable even under optimal therapy with
β-blockers and isosorbidilate. Although ambulatory 24-hr BP was not
measured, we selected patients with normal BP and it is unlikely that
all episodes of TMI were related to a marked increase of BP. It is
also unlikely that the ECG changes were artifacts as the electrodes
were correctly applied and the morphology was similar to that reported by other investigators[1-13]. Furthermore, we can exclude any
influence of hyperkalemia or other electrolyte changes known to
affect the ECG repolarization phase[20] as our patients had normal
serum values for potassium, sodium, chloride and calcium.

The association between abnormal wall motion and ST changes has
been demonstrated by experimental work[21]. Although our 22 patients
with ST-T changes had poorer left ventricular function, as shown by a
lower ejection fraction, than the 18 patients without ECG changes, it
is not possible to determine from our data whether the state of
myocardial contractility plays an important role in inducing TMI or
whether depressed myocardial contractility and ECG signs of ischemia
are merely manifestations of a primary event.

The episodes of TMI in 22/40 patients with stable and typical
angina pectoris were more frequent than suspected. In fact, of the
788 ECG-defined ischemic events (obtained during 2,885.5 hours of
DCG monitoring in these 22 patients), only 276 episodes (35%) were

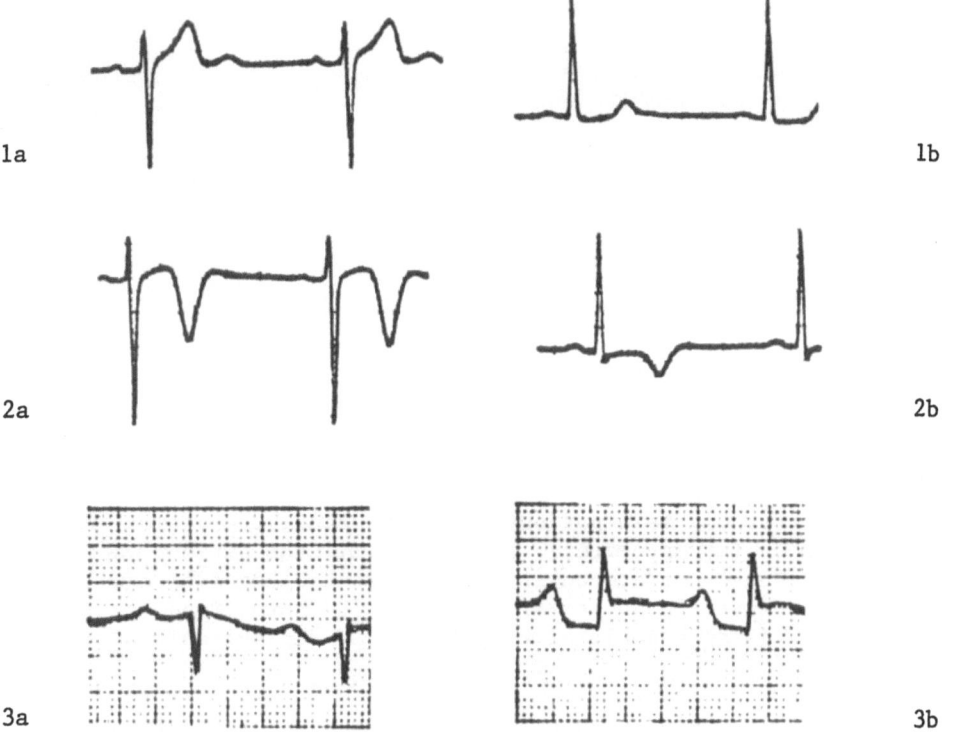

Fig. 2.   Samples 1a, 1b, 2a and 2b: 58:year old male patient. Sample
           1a (1st channel) and sample 1b (2nd channel): baselina
           (normal ECG). Samples 2a and 2b: same patient, sitting at
           home, asymptomatic. Samples 3a (1st channel) and 3b (2nd
           channel): 62-year old male patient, at work (clerk), ST
           elevation, the patient is asymptomatic.

associated with chest pain or equivalent symptoms. The asymptomatic
episodes consisted of both ST elevation or depression, and T-wave
changes. However, the duration of ECG-defined TMI was significantly
shorter in asymptomatic than symptomatic episodes. Schang and Pepine[2]
showed that 72% of the ECG ischemic events were asymptomatic, their
results being similar to ours. These authors[2] demonstrated that
these episodes were significantly reduced by nitrogylcerin. In a
previous study[4], some of us have demonstrated that the calcium-entry
blocker nifedipine was effective against resting angina pectoris and
TMI, while the β-blocker pindolol was not effective. Other investigators[6,7,9] have demonstrated the positive effect of calcium-entry
blockers in patients with variant or unstable angina pectoris. If
one assumes that coronary "spasm" plays a role in the occurrence of
TMI and spontaneous angina pectoris, then the therapy with a calcium-
entry blocker might be useful in patients with frequent TMI. At

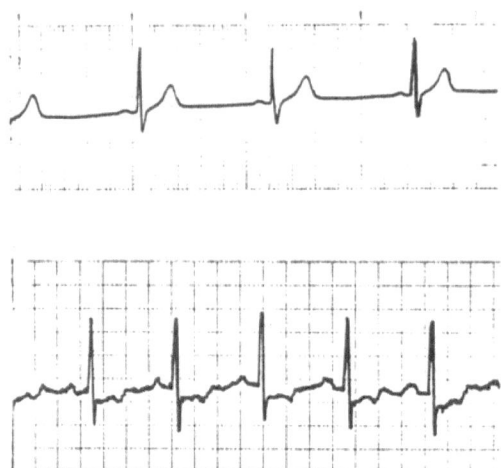

Fig. 3. 56 year old male patient. A: 6.32 a.m., the patient was
         asleep (asymptomatic). B: 6.54 a.m., the patient woke up
         because of severe chest pain, he took 0.5mg sublingual
         nitroglycerin and after 3 min the ECG normalized (as in A).
         This is an example of symptomatic ST depression.

present, however, this is an unproven hypothesis. Furthermore, our
patients were treated with effective doses of β-blockers and
isosorbidilate, and we could not interfere with their treating
physicians.

     In conclusion, our results confirm previous preliminary re-
ports[1-4,8]: the usual criterion of chest pain during daily activity
may underestimate the actual frequency of ischemic events in some
patients with CAD and stable angina pectoris. In selected patients,
adequate DCG monitoring is useful in the detection of TMI and in the
evaluation of the antianginal therapy. Additional investigation is
however necessary to determine the clinical implication of these
findings.

REFERENCES

1. S. Stern, D. Tzivoni, and Z. Stern, Diagnostic Accuracy of
    Ambulatory DCG Monitoring in Ischemic Heart Disease,
    Circulation 52:1045-1049 (1975).
2. S. J. Schang, Jr. and C. J. Pepine, Transient Asymptomatic S-T
    segment depression during daily activity, Am.J.Cardiol. 39:
    396-402 (1977).
3. P. H. Sellier, F. Proust, Ph. Delebarre, J. L. Guermonprez, P.
    Ourbak, and P. Maurice, L'enregistrement électrocardio-

graphique continu dan l'insuffisance coronaire, <u>Arch Mal Coeur</u>. 6:638-644 (1978).

4. G. Cocco, C. Strozzi, D. Chu, R. Amrein, and E Castagnoli, Therapeutic Effects of Pindolol and Nifedipine in Patients with Stable Angina Pectoris and Asymptomatic Resting Ischemia, <u>Eur.J.Cardiol</u>. 10:59-69 (1979).

5. A. P. Selwyn, K. Fox, M. Eves, D. Oakley, H. Dardie and J. Shillingford, Myocardial Ischemia in Patients with Frequent Angina Pectoris, <u>Brit.Med.J.</u> 2:1594-1596 (1978).

6. A. Maseri, S. Severi, M. De Nes, A. L'Abbate, S. Chierchia, M. Marzilli, A. M. Ballestra, O. Parodi, A. Biagini, and A. Distante, "Variant" angina: one aspect of a continuous spectrum of vasospastic myocardial ischemia, <u>Am.J.Cardiol</u>. 42:1019-1035 (1978).

7. M. Previtali, J. A. Salerno, L. Tavazzi, M. Ray, A. Medici, M. Chimienti, G. Specchia, and P. Bobba, Treatment of Angina at rest with Nifedipine: A short-term controlled study, <u>Am.J. Cardiol</u> 45:825-830 (1980).

8. G. Germanò, A. Appolloni, M. Ciaravella, A. de Zorzi, M. Fornarelli, A. De Angelis, G. Calcagnini, and V. Corsi, Proposta di standardizzazione delle metodiche nel monitoraggio ambulatoriale, Nota I: Scelta delle derivazioni, <u>Boll.Soc.Ital.Cardiol</u>. II:1307-1313 (1980).

9. A. Biagini, M. G. Mazzei, C. Carpeggiani, R. Testa, R. Antonelli, C. Michelassi, A. L'Abbate and A. Maseri: Vasospastic Ischemic Mechanism of Frequent Asymptomatic Transient ST-T changes during continuous Electrocardiographic Monitoring in Selected Unstable Angina Patients, <u>Am.Heart J.</u> 103:13-20 (1982).

10. S. M. Johnson, D. R. Mauritson, M. D. Winniford, J. T. Willerson, B. G. Firth, J. R. Cary, and L. D. Hillis, Continuous electrocardiographic monitoring in patients with unstable angina pectoris: identification of high-risk subgroup with severe coronary disease, variant angina, and/or impaired early prognosis. <u>Am.Heart J</u>. 103:4-12 (1982).

11. R. N. MacAlpin, Relation of coronary arterial spasm to sites of organic stenosis. <u>Am.J.Cardiol</u>. 46:143-153 (1980).

12. R. N. MacAlpin, Correlation of the location of coronary arterial spasm with the lead distribution of ST segment elevation during variant angina, <u>Am.Heart J</u>. 99:555-564 (1980).

13. J. J. Rozanski, N. El-Sherif, M. Kleinfeld, R. Lazzara, and R. J. Myeburg, ST segment alternans and ventricular arryhthmias in Prinzmetal's angina, <u>Circulation</u> 57 and 58 Suppl. II:239 (1978), abstract.

14. R. Haiat, C. H. Halphen, J-P Derrida, J-P Lehener, and P. Chice, Inversion transitoire d'ondes T négatives chez les coronariens. Aspect électrocardiographique inhabituel de l'ischemie myocardique spontanée. <u>Nouv.Presse méd</u>. 5:3015-3016 (1976).

15. G. Cocco and N. Rochat, Letter to the editor: Frequency response characteristics of ambulatory ECG monitoring systems and their implications for ST segment analysis. Am.Heart J. (in press).

16. C. Strozzi, G. Cocco, R. Pansini, R. Bulgarelli, A. Padula, C. Strisi, and N. Rochat, Scelta delle derivazioni nel monitoraggio con elettrocardiografia dinamica allo scopo di studiare le variazioni del segmento S-T. Convegno Naz. Elettrocardiografia Dinamica, AISC, 16-17th April (1982), Rome. Proceedings (in press).

17. L. Schamroth, "The Electrocardiography of Coronary Heart Disease," Blackwell Sci.Publ., Oxford-London-Edinburgh-Melbourne, p.136 (1976).

18. Z. Vlodaver, K. Amplatz, H. B. Burchell, and J. E. Edward, "Coronary Heart Disease. Clinical, Angiographic and Pathologic Profiles," Springer-Verlag, NY-Heidelberg-Berlin, p. 307 (1976).

19. L. Navarro, J. Cinca, G. Sanz, A. Periz, J. Magrina, and A. Betriu, Isolated T wave alternans, Am.Heart J. 3:369-374 (1978).

20. K. K. Chawla, J. Cruz, N. E. Kramer, and W. D. Towne, Electro-cardiographic changes simulating acute muocardial infarction caused by hyperkalemia: report of a patient with normal coronary arteriograms. Am.Heart J. 95:637-640 (1978).

21. C. I. Carp, Causes of S-T segment elevation, Circulation 56:498-499 (1977).

PHYSICAL ACTIVITY AND PHYSICAL FITNESS IN PRIMARY AND SECONDARY PREVENTION OF CORONARY HEART DISEASE

H. Denolin

Laboratory for Cardiac Research
Hôpital Universitaire Saint-Pierre
322, rue Haute, 1000 Bruxelles, Belgium

The specific role of sedentarity in the development of therapy of coronary heart disease (CHD) has been investigated extensively during the last 20 years.

I. Physical activity in the primary prevention of coronary disease

The few studies performed on <u>animals</u> are controversial, and their outcome different from the long-term studies on the effects of physical activity in man.

In <u>rats</u>, a better perfusion of the myocardium was demonstrated after training, without increase of a collateral circulation and without decrease in blood cholesterol or triglycerides. A possibility of a protective effect in cases of experimental infarction was suggested. A recent paper by Kramsch et al.[1] suggests that moderate exercise may prevent or retard coronary heart disease in <u>monkeys</u> submitted to an atherogenic diet.

In <u>man</u>, anatomical and clinical studies have been carried out, both retrospectively and prospectively and many critical surveys of these studies were published during the last years.

From the excellent surveys of Leon and Blackburn[2], Froelicher[3,4] and others, it appears that the results are controversial, but that the studies in favor of a protective effect of physical activity are more numerous than studies in which a sedentary life style is considered to be of no importance in the development of ischemic heart disease.

In histo-pathological studies, it was found that high level sportsmen – and especially Marathon runners – have large coronary arteries, with less atherosclerosis; but recently, this concept was questioned and in any case, it is difficult to use these anatomical data, based only on a very few autopsies[5,6].

The following critical remarks concerning clinical studies, should be considered:

(a) Vocational and leisure time activities are not allocated in a randomized way, and self-selection must play a role in the choice of vocation or sport.
(b) Other risk factors and social circumstances are often over-looked, particularly in the retrospective studies.
(c) A genetic factor is possible. In a paper by Siegell[7], 18% of cardiac events were observed in the parents of marathon runners, as against 34% in the control group.
(d) Changes of jobs are frequent, and changes in activity are very frequent after a few years in high level athletes.
(e) The true level of vocational activity is difficult to assess, as in leisure activities, and the questionnaires reflect only the activities of the last few months or years.
(f) The definition of ischemic heart disease may vary from one study to another.
(g) The cause of death is not always exactly known.

Finally, the physical fitness is not always considered; this physical fitness may vary greatly between subjects at the beginning of the studies, may effect the vocational or sport performance, or may reflect a genetic background influencing the attitudes of the subjects and the development of the disease.

The following two studies are in favor of a protective effect against the development of an IHD:

Brunner et al.,[8] have studied 5,288 men and 5,299 women in the medium age (40-64 years) working in Kibbutzim, living under similar conditions and with the same nutritional regimen. Observations were carried out over a period of 15 years and it was found that coronary events appear 3 times more in the sedentary subjects than in the more active.

Other well known studies were published by Paffenberger[9]. In longhorsemen, the more active develop IHD 3 times less frequently than the sedentary subjects; the difference is especially striking in relation to sudden death. In an other paper[10], the same author demonstrates that the risk of a first heart attack is inversely related to energy expenditure in Harvard male alumni aged 35-74 years.

In Italy, Menotti and Monti[11] found that in 172,000 men the mortality for CHD was statistically higher in sedentary and moderate than in heavy workers.

While many studies of this type are published in the literature, presenting sedentarity as a risk factor, other studies were not able to confirm these conclusions.

As an example, let us quote the study of Rosenman[12], from which it appears that, in a group of 2,065 white males, aged 35-59 years, the development of an ischemic complication is related to age, cholesterol, systolic blood pressure and cigarettes smoking, but not to the level or type of vocational activity or to the total activity, including leisure activity.

In Finland, Kallio[13] found no correlation between cardiovascular death and the level of physical activity, but observed that an increase in leisure activity is associated with a lower cigarettes consumption, a reduction in obesity, a lower arterial pressure and a lower cholesterol level.

Considering the possible importance of physical fitness related to a genetic factor or as a result of physical activity, we started recently in Belgium with a study attempting to correlate an actual objective physical measure of physical capacity with a physical activity classification[14,15]. 1,473 men of 40-55 years were eligible to enter a 5 years follow-up, following an exercise test on the bicycle ergometer, a clinical and radiological examination and a questionnaire on activities during the proceeding 12 months period. The available information demonstrates that the leisure time activity score was significantly related to physical fitness and independently, that the job physical activity was also significantly related to physical fitness. The data illustrates an overall low energy expenditure profile of middle age normal men and a low order relationship between physical activity pattern and physical fitness. The results of this study are not yet known, and the present conclusion is:

If in this study, fitness levels are shown to be related to CHD incidence, this could reflect differences other than in physical activity levels; furthermore, if job or leisure time activities are shown to be related to CHD, this could reflect differences other than fitness[15].

In the Göteborg study[16] the incidence of MI during a follow-up period of 9 years was related to the maximal work capacity. The problem remains to decide whether physical fitness is related to leisure time or job activity or to a constitutional factor.

In the study of Gyntelberg[17] a follow-up study of 5,249 males, aged 40-59, showed after 7 years that the incidence of MI was related to the level of physical fitness at the time of entry into the study, as determined by the indirect measurement of maximal oxygen uptake (Astrand Nomogram); only physical activity inducing a physical training effect is related to a decreased risk.

The influence of job or leisure time activities on the risk factors is difficult to evaluate, and the analysis of all the numerous studies available on the relationship between physical activity or physical fitness and the development of a CHI remain inconclusive should indeed a positive effect exist, it still remains questionable whether this is a direct effect or an effect through reduction of other risk factors.

These questions raise the problem of the possible mechanisms through which physical activity could act (Table 1). Many suggestions were presented. The physiological effect of training, on heart, lungs, muscles, blood and the psychological behaviour have been studied extensively, but we have to define the mechanism through which physical activity could have a preventive effect. The hypothesis "à la mode" is the effect of physical activity on HDL lipoproteins. A protective effect of HDL was advocated in some studies and some studies demonstrate an increase in HDL through physical activity, but unfortunately this was not supported by all the studies[18,19,20]. The effect on triglycerides is controversial and, the total cholesterol is probably not affected by exercise.

Finally, if we accept a favorable effect of exercise in primary prevention, what should be the level of the preventive activity? To what extent should leisure time activity be added to vocational activity. What type of exercise? What intensity? What duration? Should this be a continuous or discontinuous program? Up to what age are such programes useful? We have still no clear answer to these questions.

In summary, we have no adequate answer to the following questions

(a) Does physical activity have a preventive effect in the development of CHD?
(b) If yes, should we consider physical activity or working capacity as the preventive factor.
(c) How does the physical activity exert its benefit?
(d) What is the optimal program?

Of course, the physiological and psychological benefits of physical activity are well known, and regular activity is at least able to improve the "joie de vivre."

Table 1. Possible effects of physical activity

Development of coronary arteries
Growth of capillaries
Improvement of contractility
Decrease in Systemic arterial pressure
Reduction of Obesity
Reduction of adrenergic activity
Reduction of risk factors (Tobacco)
Changes in blood lipids
Reduction in Glucose tolerance
Modification of coagulation and fibrinolysis
Electrophysiologic modifications (arrhythmias)
Psychologic changes

II. Physical activity in the secondary prevention of coronary disease

The idea that some physical activity could be beneficial in
CHD was already suggested by Heberden. But in fact it is only after
1944 and the papers of Levine and Lown that it was progressively
accepted, first that a reduction of immobilization could be beneficial, and then that physical training could have a secondary preventive effect. Today, patients are mobilized very early after a
MI, leave the hospital after 10 to 15 days and most of them are
submitted to a supervised training after 3 to 8 weeks. During the
last years, it was also demonstrated that the psycho-social approach
and the secondary prevention - by correction of the risk factors and
the use of drugs - could be of great importance, at least for a
short time.

We know the effects of physical training on the physical condition of the patients after MI: increase in maximal oxygen consumption, of 20 to 30%, increase in the functional capacity, use of
a lesser part of the maximum during job activity. It was accepted
that the improvement in the physical capacity is probably related
more to a better extraction of oxygen from the blood at the periphery
than to a change in the myocardial function[21]. But, recently, it
appears that at least in some cases the perfusion of the myocardium
is improved after training, may be by an increase in collateral
circulation, and it was also said that the ejection fraction could
be increased, at equivalent submaximal level of exercise, in the
trained patients[22,23].

But what are the results of the cardiac rehabilitation program
at long term, on morbidity and mortality? The results are still
controversial.

In the old studies, the results of physical training were evaluated without a control group; in a few more recent studies, a program of supervised training was designed for each patient randomized to intervention. The level of training is usually based on achieving submaximal levels of stress as determined either by a prior exercise test or by age related pulse rate; the program is adminstered in a clinic, with 2 to 4 sessions of training per week, for 20 to 60 minutes. In the control group, in general, the patients received regular medical care with no special emphasis on exercise.

In these randomized studies, the total number of patients varied between 298 and 733. The patients enter the training program on hospital discharge or later: to 12 months or more in the national exercise heart disease program (NEHDP)[24]. In all the studies, the mean age of the participants was around 50. The follow-up is from 1 to 4 years. The compliance to the exercise program was generally poor.

From the analysis of May and coworkers[25], it appears that none of the trials showed a statistically significant difference in total mortality between the exercise and the control group; all, except one, however, showed a positive trend favoring the exercise group. There appeared to be little effect on the incidence of reinfarction, except in one group.

In the NEHDP, a reduction of 37% in mortality was observed in the exercise group: the mortality rate was 7.3% in the control group and 4.6% in the exercise group. These results do not indicate benefit concerning cardiovascular morbidity. The numbers of patients are low (651) and the patients enter the program very late[24].

In the Ontario Exercise-Heart collaborative study[26], it appears that the drop out was very high, in the control group as well as in the exercise group, and that the rate of recurrence is not significantly different between the two groups.

The European Office of WHO was well aware of these problems many years ago, and started with a coordinated study on the effects of rehabilitation and comprehensive secondary prevention[27]. The total number of patients is 3,118 recorded from 19 European centers; patients were under the age of 65, with definite MI, treated in a hospital during the acute phase. The intervention measures were to be applied according to the best knowledge in each individual center. The follow-up period was 3 years. The main endpoints were death and morbidity. Physical working capacity, changes in the quality of life and reduction of risk factors were used as additional criteria.

Considering the endpoints, pooling of all the data is quite impossible, because of local attitudes and other reasons. If one considers the 17 centers with a sufficient number of patients in the treated and control group, 13 have a mortality experience which favor the rehabilitation group. The difference in reinfarction, in the 2 groups, is not significative. In most of the centers, the physical capacity is improved.

The local differences in results are very important, and demonstrate how difficult the long term evaluation is. In the group of patients collected by Kallio[28] in Finland for this WHO study, 375 patients were included. The total mortality and the cardiac mortality are lower in the intervention group, suggesting that patients with MI would benefit from an organized comprehensive rehabilitation.

But, in Belgium, in one of the centers participating in the study, the mortality in the control group was 0% after 3 years, against 6% in the rehabilitation group. This demonstrates again the difficulties of such studies, related to the insufficient number of patients, the local differences in methodology, etc.

So, the effects of a rehabilitation program – including physical training at least during several weeks after the MI – remain difficult to evaluate, and this is probably related to many reasons: low number of subjects, short follow up, low compliance to the program, drop out, mode of selection in the different centers, physiological criteria for admission (including left ventricular function,) and probably the very important problem of drop in, that is contamination of the control group by the new therapeutic attitude.

Considering those difficulties, a conclusion is proposed by Roy Shepard[29]: "Pessimist might conclude that a controlled study to determine the influence of exercise upon the rate of reinfarction is logistically impossible ... Optimists might attempt to meet the requirement of large groups by pooling data from various controlled studies, ignoring obvious differences of philosophy, and practice."

Conclusion

In the case of both primary and secondary prevention, we do not know whether we add years to life by prescribing physical activity, but it seems at least that we add life to years.

REFERENCES

1. D.M. Kramsch, A.J. Aspen, B.M. Abramowicz, T. Kreimendahl, and W.B. Hood, Reduction of Coronary atherosclerosis by moderate conditioning exercise in monkeys an atherogenic diet, New Eng. J.Med. 305:1483 (1981).
2. A.S. Leon and H. Blackburn, The relationship of physical activity to coronary heart disease and life expectancy, Ann.Acad. Sci.NY 301:561 (1977).

3. V.F. Froelicher and A. Oberman, Analysis of epidemiologic studies
    of physical inactivity as risk factor for coronary artery disease,
    Prog.Cardiovasc.Dis. 15:41 (1972).
4. V.F. Froelicher, A. Battler, and M.D. McKirnan, Physical Activity
    and Coronary Heart Disease, Cardiology. 65:153 (1980).
5. T.D. Noakes, L.H. Opie, A.G. Rose, and P.H.T. Kleynhans, Autopsy-
    proved coronary atherosclerosis in marathon runners, N.Engl.J.
    Med. 301:86 (1979).
6. R.W. Asay and W.V.R. Vieweg, Severe coronary atherosclerosis in
    a runner: an exception to the rule, J.Cardiac Rehab. 1:413
    (1981).
7. A.J. Siegel, C.H. Hennekens, B. Rosner, and L.K. Karison,
    Paternal history of coronary-heart disease reported by marathon
    runners, N.Engl.J.Med. 301:90 (1979).
8. D. Brunner, G. Maneis, M. Modan, and S. Levin, Physical activity
    at work and the incidence of myocardial infarction, angina
    pectoris and death due to ischemic heart disease. An epidemio-
    logical study in Israeli Collective settlements (Kibbutzin).
    J.Chronic.Dis. 27:217 (1974).
9. R.S. Paffenberger, Counter currents of physical activity and
    heart attack trends, in: "Proceedings of the Conference on the
    decline in coronary heart disease mortality," US Department of
    Health Education, p.298 (1979).
10. R.S. Paffenberger, A. Wing, and R.J. Hyde, Physical activity as
    an index of heart attack risk in college alumni, Am.J.Epidemiol.
    108:161 (1978).
11. A. Menotti and M. Monti, Attività fisica responsabilità
    lavorative quali fattori di rischio di cardiopatia coronarica
    fatale, G.Ital.Cardiol. 9:668 (1979).
12. R.H. Rosenman, R.D. Bawol and B. Oscherwitz, A 4-years prospec-
    tive study of the relationship of different habitual vocational
    activity to risk and incidence of ischemic heart disease in
    volunteer male federal employees, Ann.Acad.Sci.NY 301:627 (1977)
13. V. Kallio, Experience of prevention by habitual physical activity
    in: "Sports Cardiology," eds., T. Lubich and A. Venerando, p.809,
    Bologne, Aubo Gaggi, (1980).
14. J. Sobolski, G. De Backer, S. Degre, M. Kornitzer, and H. Denolin
    Physical activity, physical fitness and cardiovascular diseases:
    design of a prospective epidemiologic study, Cardiology 67:38
    (1981).
15. G. De Backer, M. Kornitzer, J. Sobolski, M. Dramaix, S. Degre,
    M. de Marneffe, and H. Denolin, Physical activity and physical
    fitness levels in Belgian males aged 40-55 years, Cardiology
    67:110 (1981).
16. L. Wilhelmsen, J. Bjure, B. Ekström-Jodal, M. Aurell, G. Grimby,
    K. Svärdsudd, G. Tibblin, and H. Wedel, Nine years follow-up of
    a maximal exercise test in a random population sample of middle-
    aged men, Cardiology 68:(Suppl.2)1 (1981).
17. F. Gyntelberg, L. Lauridsen, and K. Schubell, Physical fitness
    and risk of myocardial infarction in Copenhagen males aged 40-
    59, Scad.J.Work Environ.Health 6:170 (1980).

18. R.H. Dressendorfer, C.H.E. Wàde, C. Hornick, and G.C. Timmis,
    High density lipoprotein-cholesterol in marathon runners during
    a 20 day road race, JAMA 247:1715 (1982).
19. K.D. Brownell, P.S. Bachorik, and R.S. Ayerle, Changes in plasma
    lipid and lipoprotein levels in men and women after a program
    of moderate exercise, Circulation 65:477 (1982).
20. T. Gordon, W.P. Castelli, M.C. Hjortland, W.B. Kannel, and
    Th.R. Dawber, High density lipoprotein as a protective factor
    against coronary heart disease, Am.J.Cardiol. 62:707 (1977).
21. S. Degre, Effets physiologiques de l'entraînement musculaire
    chez le malade atteint d'infarctus du myocarde et de pneumo-
    pathie chronique. Editions de l'Université de Bruxelles. (1975).
22. K. Witztum, R. Slutsky, and W. Ashburn, Radionuclide perfusion
    images before and after cardiac rehabilitation, Aviation, Space
    and Environ.Med. 51:892 (1980).
23. D. Jensen, J.E. Atwood, V. Froelicher, D. McKornan, A. Battler,
    W. Ashburn, and J. Ross, Improvement in ventricular function
    during exercise studied with radionuclide ventriculography
    after cardiac rehabilitation, Am.J.Cardiol. 46:770 (1980).
24. L.W. Shaw, Effects of a prescribed supervised exercise program
    on mortality and cardiovascular morbidity in patients after a
    myocardial infarction. The National Exercise and Heart Disease
    Project, Am.J.Cardiol. 48:39 (1981).
25. G.S. May, K.A. Eberlein, C.D. Furberg, E.R. Passamani, and D.L.
    de Mets, Secondary prevention after myocardial infarction: a
    review of long-term trials, Progress in C.V.D. 24:331 (1982).
26. T. Kavanagh, Evidence to date for the beneficial effect of
    exercise following myocardial infarction, in: "Controversies
    in Cardiac Rehabilitation," eds., Mathes and Halhuber, Springer-
    Verlag, Berlin (1982).
27. Who, Study on the effects of rehabilitation and comprehensive
    secondary prevention in patients after myocardial infarction,
    Copenhague. To be published.
28. V. Kallio, Evaluation of earlier Studies: Europe, in: "Physical
    conditioning and cardiovascular rehabilitation," eds., Cohen,
    Mock and Ringqvist, John Wiley and Sons, New York (1981).
29. R.J. Shepard, Evaluation of earlier studies: Canada, in:
    "Physical conditioning and cardiovascular rehabilitation,"
    eds., Cohen, Mock and Ringqvist, John Wiley and Sons, New York
    (1981).

THE SECONDARY PREVENTIVE EFFECT OF COMPREHENSIVE

CORONARY CARE (C.C.C.)

Jan J. Kellermann

The Hermann Mayer Cardiac Rehabilitation Institute
Chaim Sheba Medical Center
Sackler School of Medicine
Tel Hashomer 52621, Israel

Till now there is a lack of scientific evidence on the secondary preventive effect of comprehensive intervention programs in patients with CHD. A cynical approach to this problem may always be suspicious of weakness and of scientific arrogancy. Nonetheless, I shall try to present the problem as objectively and unbiased as possible. In order to analyse the effectivity of programs implemented in the treatment of patients with CHD, it seems a necessity to discuss first of all the nature of prevention on coronary disease in general. It is common knowledge that atherosclerosis of the coronary vessels is considered to be the underlaying condition of coronary artery disease. Moreover, this disease has proved to be of multifactorial origin. While the notion that atherosclerosis is a reversible process was suggested already 60 years ago, a regression of these processes has been demonstrated only in animals. In humans such a reversibility could be shown only incompletely and it appeared that there exists many problems of interpretation. It has to be considered that the reversions of hyperlipidemic conditions may cause a regression of the atherosclerotic state. In humans such a regression may eventually be caused by a multiplicity of factors and not only by hypercholesterolemia, as in the experimental atherosclerosis induced in animals. These observations must be remembered before any conclusions can be drawn as to the effect of preventive measures in individuals without signs and symptoms of coronary atherosclerosis[1].

Of late, a decline in CHD death rates has been reported especially in the USA, Finland, New Zealand, Norway, Canada, Israel and a number of other countries. Epidemiologists concluded that this decline may be due to a change in risk factor distributions

and habits of certain populations. The North Karelia Project has
shown that, using a multiple logistic analysis for the three main
risk factors such as smoking, serum cholesterol and systolic blood
pressure, there was a significant net reduction of 17.4% in men and
11.5% in women in the estimated coronary risk in North Karelia.
From 1972-1977 the decline in the incidence rate of acute myocardial
infarction was 16.7% in men and 10.2% in women[2].

It is known that the mortality of CHD in Finland was one of the
highest in the world and in the line of constructive criticism one
may argue that the epidemic of CHD has reached in some countries a
culminating point. We definitely do not concur with Ivan Illich,
who says in his "Limits to Medicine" that - "The study of the evolution of disease patterns provides evidence that during the last
century doctors have affected epidemics no more profoundly than the
priests during earlier times". On the other hand, his statement
that "epidemics came and went imprecated by both but touched by
neither" cannot simply be ignored, because this latter statement was
based on observations with mortality rates from tuberculosis and
other infectious disease in the past 150 years. Therefore, we should
be careful, sceptic and logical when analysing the reasons for the
decline of CHD mortality in the past decade.

One must take into account that atherosclerosis of the coronary
arteries is a degenerative process developing throughout the years,
starting probably at an early age and being influenced by a great
number of various factors. The reason for the clinical outburst of
the disease, i.e. myocardial infarction, angina pectoris, sudden
death, is still not quite clear. It seems therefore rather bold to
conclude that the change of habits and risk factors in a time period
of less than 10 years will affect significantly morbidity and mortality.

Together with the reported decline in CHD mortality between
1971-80, there was an unprecedented increase in costs for the treatment of heart disease in the same period of time. In 1971, 3.7
billion dollars were spent in the USA for drugs, bypass surgery,
pace-makers and diagnostic procedures such as coronary angiography
and others. In 1982 the expenditure for diagnostic and therapeutic
management of heart disease in the USA was 39.3 billion dollars[3].

It seems to us that there may be a link between earlier, better
and more comprehensive care and the decline in the death rate. In
most of the countries evidence is lacking that the decrease in mortality was accompanied by a decrease in morbidity. In some countries
there even seems to be an increase in morbidity during the same
period of time.

Concluding this part, we should like to state that, while there
is a general consent as to the influence of major risk factors (such

as hyperlipidemia, hypertension and smoking) on morbidity and on mortality of CHD, the notion that the change of habits and the effective control of major risk factors for a relatively short period of time will result in a dramatic decrease of a multifactorial disease seems to me somewhat obscure.

We are not trying to diminish the importance of extensive epi-demiologic research, especially during the past two decades, but we would like to be cautious about any statements and conclusions indi-cating that primary prevention and especially preventive strategies implemented during a couple of years, cause a decline in mortality of CHD. We still have to find out whether we are not facing a "natural decline" of an epidemic.

CARDIAC REHABILITATION AS A SECONDARY PREVENTIVE MEASURE

The initiation of comprehensive rehabilitation programs in patients with coronary heart disease still remains controversial as to the prevention of mortality and morbidity[4].

For the past decade many scientists, professional national and international institutions stressed the urgent need of a well con-trolled, randomized trial in order to find an answer to this latter question. Despite the evidence that such trials are very expensive and their cost benefit effectiveness has yet to be established, nobody was prepared to compromise with a more flexible and less rigid scientific approach.

FEASIBILITY OF RANDOMIZED TRIALS

Requirements – Problems – Biases[5]

The following aspects must be taken into consideration:

a)   It is desirable to conduct a randomized trial taking into con-sideration all the aspects of a pure scientific procedure. In order to perform such a trial we have to face a number of very difficult problems, such as exact dosage, patient's compliance and contaminations.
Furthermore, endpoints must be clearly defined because there may be a number of detection biases, identification difficulties and a lack of unity for the classification of symptoms and of signs.

b)   The other problems consist of the so-called preintervention base line state. When dealing with a multifactorial disease such as coronary heart disease, one must be especially aware to assess constantly not only the severity of symptoms and the symptoms combination, but also the rate of progress. As a consequence of cointerventions, changes in the titration of dosage may be necessary.

c)   In coronary heart disease with a pathogenesis of atherosclerosis of coronary vessels, a great number of risk factors may influence the natural course of the disease[6]. The problem of "the therapeutic base line state" may present an obstacle almost impossible to overcome. The following risk factors may eventually influence prognosis, either by interaction or by sudden uncontrollable events[1]: heredity, impaired lipid metabolism, diabetes, hypertension, smoking habits, obesity, physical inactivity, personality traits and behavior patterns, psychosocial stresses, ethnic-cultural backgrounds.

d)   Supervised rehabilitation programs which include systematic clinical reassessments, modification of risk factors and habits, psychological and vocational counselling, treatment of hypertension, arrhythmias, diabetes and angina pectoris and a physical training program, may cause difficult problems of cointerventions in the individual patient. Comprehensive coronary care may influence directly the outcome of events, either as a consequence of a single therapeutic procedure, or as a result of comprehensive therapy. Prognosis may therefore be influenced in all these patients who are reacting beneficially to therapy and who are eventually preselected due to a consequent adherence to long term programs of this kind.
     It is mandatory that any results of randomized studies can be taken into consideration only when their duraction extends to a large number of years. In most of the randomized studies undertaken to date, the ambitions to achieve hard endpoint criterions, proved to be a "mission impossible". What we have to accept is the fact that a great number of biases cannot be eliminated when intervening in different groups of patients with CHD.

RANDOMIZED STUDIES

     The U.S. National Exercise Project[7] reported the results after 3 years follow up. It was found that in the exercise training group there was a significant lower death rate from myocardial infarction when compared to the control group. The accumulated 3 years total mortality rate was 7.3% for the control group and 4.6% for the intervention group. But it must be stated that, while there was an adherency of 73% to a minimum training regime in the intervention group, 31% of the control group was engaged in some regular physical activity. The conclusion of these findings is that, not only drop-out rates may seriously influence the outcome of such studies, but also the drop-in rate.

The WHO Multicenter Trial[8] reported that the pooling of data
was complicated because of different local attitudes. Big inter-
center differences were found concerning mortality. Out of 17
centers, 13 had lower mortality rates in the rehabilitation groups.
But it should be stated that the follow up was only for three years.
It was concluded that the study failed to give an unequivocal answer
to the main questions asked. Neither mortality nor morbidity, nor
the softer end points like return to work, demonstrated consistent
benefits from systematic rehabilitation. Kallio[9] reported from
Finland that in his study, which was part of the WHO project, the
total and cardiac mortality was lower in the intervention group.
On the other hand, in one of the Belgian studies, the mortality in
the control group was 0% as against 6% in the rehabilitation group.
The follow up of both studies was three years[10].

The outcome of these randomized trials does not yet permit
definitive statements as to the impact of cardiac rehabilitation
programs on secondary prevention. So far we have failed to prove
that our hypotheses are true. It may well be that the reason why
we are unable to prove them is the wrong questions we are asking.
It may be necessary to rephrase the designs of our studies and
create new testing devices.

Of late, Froelicher and co-workers[11] have initiated a randomized
trial on the influence of physical training on left ventricular
function. In preliminary findings they found that in selected
patients with CHD, the ejection fraction can be enhanced as a result
of effective physical training. The outcome of this trial may prove
to be of utmost importance, because it should be possible to reach
by deduction conclusions on the prognostic influence of physical
training, taking into consideration that the left ventricular function
is being considered to be a very important prognostic sign.

NON-RANDOMIZED TRIALS

A fairly large number of non-randomized studies have been
carried out in the past 20 years. As these studies were widely
published, we deem it unnecessary to repeat their outcome. It seems
indisputable that these studies have actually paved the way for the
discipline of cardiac rehabilitation, which has since then gained
wide recognition and popularity. Nonetheless, the outcome of these
studies, indicating a trend of decreased mortality in intervention
groups as compared to non-randomized control groups, does not constitute a scientific conclusion as to the secondary preventive
effect of rehabilitation programs.

A number of studies initiated by us since 1962, comprised of
patients after myocardial infarction with and without angina pectoris
who underwent continuous up to 10 years supervised rehabilitation

programs, showed an annual mortality of 0.8% from cardiac causes,
while the control groups had death rates between 2.3 to 2.7% [12].
In a recently conducted study, results after a 7.6 years follow up
in 197 patients after myocardial infarction with a mean age 49.5 ±
4.1 years were summarized. The program included risk factor modifications, physical training and drug therapy. 11.7% of the patients
underwent aorta coronary bypass surgery.

The results of our study indicated that the physical work
capacity increased at a level p<0.001, the double product and heart
rate decreased for a given work task (p<0.001), the oxygen pulse
beat/min. increased for a given load (p<0.01). Total cholesterol
decreased (p<0.001) and so did the resting systolic and diastolic
blood pressure (p<0.001). The mean cardiac mortality was 0.65% per
year. As can be seen from Table 1, only 19.3% underwent physical
training, while 69% received additional drug therapy such as beta
blockers (33.5%), calcium antagonists (22.5%), antiarrhythmic therapy
(31.8%), nitrates (30.8%), digitalis (3.5%). A number of patients
received more than one compound[13].

It is obvious that no conclusions can be drawn as to the effect
of a physical training program per sé, on the outcome of this study.
A multifactorial intervention is tantamount and the implementation
of a number of therapeutic procedures as well as the placebo effect
of the "institutional frame" being applied in these kind of programs,
make it impossible to analyse scientifically whether or not the low
mortality is due to the intervention as such, in a preselected group
of patients. This preselection is almost inavoidable because of
the fact that in order to maintain adherence, these patients must
have a positive attitude to undergo supervised and medically controlled regimes for a prolonged period of time and have to be aware
of health hygiene (our mean drop-out rate was 3.3% per year)[14].

Table 1. Mode of Intervention

	N	%
Exercise only	38	19.3
Exercise + Drugs	136	69.0
Exercise + Surgery	7	3.6
Exercise + Surgery + Drugs	16	8.1
Total	197	100%

CONCLUSIONS - THE CONCEPT OF COMPREHENSIVE
CORONARY CARE (C.C.C.)

The restoration of the patients physiological, psychological
and social capabilities is aimed to improve the quality of survival.
It is tantamount to treat a multifactorial disease by a multifactorial
approach. Until the end of the sixties rehabilitation was identified
with physiotherapy and physical training. It was only during the
last decade that a comprehensive approach was systematically accepted,
mainly that physical training is only one therapeutic modality which
can be applied in selected groups of coronary patients, and that
rehabilitation procedures must include risk factor modifications,
change in life habits and other conservative and/or surgical therapeutic measures. Furthermore, one of the most important clues of
C.C.C. is a systematic follow up. By means of non-invasive techniques
such as exercise testing, echocardiography, nuclear imaging and
others, it is possible to pay attention to any change in clinical
and functional conditions and the dynamics of the disease. Such an
early detection of signs of deterioration may enable a timely initiation of effective therapeutic measures.

Scientific evidence is missing that continuous rehabilitation
programs have an effect on secondary prevention. Randomized trials
have so far failed to produce meaningful results concerning mortality, while non-randomized control trials have shown a lower mortality in the intervention group. Rehabilitation programs have to
be considered as an integral part of C.C.C. The early initiation
of such comprehensive interventions in the acute, convalescent and
maintenance phase after myocardial infarction, may have contributed
to the decline in death rates in some countries. Quality of survival can certainly be enhanced by comprehensive rehabilitation,
whether or not the survival itself can be influenced still remains
to be demonstrated.

REFERENCES

1. J. J. Kellermann, Can we improve the prognosis of coronary
       heart disease, In P. Mathes and M. J. Halhuber (eds): Current
       Problems in Cardiac Rehabilitation, Springer - Verlag, Berlin
       - Heidelberg - New York, pp. 123-129 (1982).
2. Community control of cardiovascular disease, The North Karelia
       Project, World Health Organization, Copenhagen (1982).
3. America's $39 Billion Heart Business in US News & World Report,
       March 15 (1982).
4. J. J. Kellermann and H. Denolin, Critical Evaluation of Cardiac
       Rehabilitation, Bib. Cardiol., 36, S. Karger AG, Basel -
       New York (1977).
5. A. R. Feinstein, Clinical Biostatistics, C.V. Mosby Co., St.
       Louis, (1977).

6. J. J. Kellermann, Cardiac Rehabilitation as a secondary pre-
   ventive measure. Endpoints: The Logic of Desirability and
   Availability in J. J. Kellermann (ed.) Comprehensive Cardiac
   Rehabilitation, Adv. Cardiol., vol. 31, pp. 134-137, Karger,
   Basel (1982).

7. L. W. Shaw, Effects of a prescribed supervised exercise program
   on mortality and cardiovascular morbidity in patients after
   myocardial infarction, The National Exercise and Heart
   Disease Project, Am. J. of Cardiology, 48:39-46 (1981).

8. G. Lamm, H. Denolin, D. Dorrosiev and Z. Pisa, Rehabilitation
   and secondary prevention of patients after myocardial infarction, see ref. 6, pp. 107-111 (1982).

9. V. Kallio, Rehabilitation programs as secondary prevention: A
   community approach, see ref. 6, pp. 120-128 (1982).

10. H. Denolin, Present and future of cardiac rehabilitation, see
    ref. 6, pp. 102-106 (1982).

11. V. F. Froelicher, D. Jensen, J. E. Atwood, D. McKirnan, K.
    Gerber, R. Slutsky, A. Battler, W. Ashburn and J. Ross,
    Cardiac rehabilitation: Evidence for improvement in myo-
    cardial perfusion and function, Arch. Phys. Med. Rehab.,
    61:517-522 (1980).

12. J. J. Kellermann, Rehabilitation of patients with coronary
    heart disease, Prog. in Cardiovascular Diseases, vol. 17:4,
    pp. 303-328 (1975).

13. J. J. Kellermann, E. Ben Ari, Y. Drory and M. Hayet, Unpublished
    data (1982).

14. J. J. Kellermann, Cardiac rehabilitation: Reminiscences, inter-
    national variations, experiences, Journal of Cardiac Rehabili-
    tation, 1:43-50 (1981).

CORONARY ARTERY SURGERY:

RANDOMIZED CONTROLLED TRIALS INDISPENSABLE

David H. Spodick

University of Massachusetts, Medical School
Division of Cardiology
St. Vincent Hospital
Worcester, Massachusetts 01604

Coronary artery surgery is one of a number of treatments of coronary artery disease and its manifestations. As with any treatment, the standards for judging its value should be equal to the standards for all other treatments. Since even under the best circumstances, surgery and anesthesia themselves impose risks on patients, however small, the standards should be no less stringent, when feasible, than for other kinds of treatment. That means that to judge the effects of this particular treatment, they must be compared on an appropriate basis with those of alternate treatment in truly comparable patients, keeping in mind - and making provision for - changes over time, changes in non-surgical treatment, changes in the "natural history" of coronary disease and improvements in surgical and allied techniques and skill.

It has been repeatedly shown that the optimum method of assessing most treatments is by appropriately designed prospective randomized controlled clinical trials. It has also been shown that such trials are quite feasible for surgical treatments, including coronary artery surgery. Thus, the case for randomized controlled trials (RCTs) of coronary surgery rests on the same three bases as for other treatments. These bases are scientific, behavioral and ethical.

THE SCIENTIFIC CASE

The scientific case for RCTs of coronary surgery rests squarely
on the power of randomization to assign comparable patients to either
surgical or nonsurgical treatment. This minimizes or eliminates biases
in selecting patients – a prime source of error in judging many treatments, notably those involving surgery. Decades of experience document enthusiastic reports of "success" of numerous treatments which
finally were invalidated either by traditional trial-and-error or by
appropriately designed RCTs. Many of these treatments were conceived
on the basis of what was considered "obvious" correction of an anatomic
defect, e.g. surgery for coronary disease – pericardiopexy and poudrage,
Vineberg implants, omentopexy, internal mammary ligation and simple
endarterectomy; other kinds of operations – sympathectomy for hypertension, gastropexy for "gastroptosis", nephropexy for "nephroptosis",
glomectomy for asthma, prophylactic portacaval shunt and colon bypass
for hepatic encephalopathy. All the foregoing were enthusiastically
proclaimed as effective cures and were used in many thousands of
patients, but finally were discarded. Some of the reasoning behind
such treatments was that the apparent structural changes would
"obviously" have the desired effect.

"Obvious" Solutions

The fate of "obvious" cures is one of the less glorious pages
in medical history. Examples of the alluring "obvious" answers for
complicated problems can be dealt with quite briefly:

1. Radical breast resection for carcinoma. We now know that this
    is often unnecessary and, in fact, in addition to the mutilating
    operation itself, resection of axillary nodes and radiation may
    not be successful.

2. High-dose insulin for diabetic ketoacidosis. This was accepted
    as axiomatic. We know now that this problem is frequently solved
    with quite low doses of insulin.

3. Antacids for peptic ulcer are another "obvious" solution and were
    so until it was shown that there is an increased "acid tide" that
    excessively lowers pH following initial relief of symptoms by
    antacids.

4. High-dose oxygen for premature infants. This was not obviously
    detrimental until it was shown in formal investigations (planned
    and performed despite much opposition) that this was the cause
    of retrolental fibroplasia.

5. Sympathectomy for hypertension. The placebo effect of this type
    of surgery was never considered; it is no longer used, fortunately,
    because of effective medical treatment.

6. <u>Estrogen therapy for prostatic carcinoma</u>. Despite the effects
   on the carcinoma there was an increase in coronary disease.

7. <u>It has been considered that increased myocardial blood flow would
   "obviously" improve patients with coronary disease</u>. However, di-
   pyramidole, originally touted as an "oxygen deliverer" to the
   myocardium, does increase subendocardial blood flow but does not
   deliver oxygen to subendocardial muscle. In contrast, nitroglycerin
   does not necessarily increase subendocardial blood flow but causes
   subendocardial tissue oxygen to rise.

8. <u>Reducing plasma cholesterol</u> was considered as obviously improving
   prognosis in patients with arterial disease. However, the WHO
   study showed that, despite significant reduction in cholesterol
   by clofibrate, patients who had taken that agent had a higher
   overall death rate, including coronary death rate, than those
   taking placebo, whose cholesterol did not change.

"Obvious" solutions are often like a dead herring in the moon-
light - it shines, but it smells.

Enthusiasm in Reports of Trials

When one examines the work of investigators who have not done a
randomized controlled trial one finds that their reports are nearly
always enthusiastic - both for treatments that later prove to be effective and for those ultimately shown to be worthless. In contrast,
the level of enthusiasm in reports of well controlled trials (even
when the agent tested is effective) is nearly always much less. Indeed, the level of enthusiasm tends to be in inverse proportion to
the scientific quality of the investigation.

Historical "Controls"

In uncontrolled investigations there is either a general comparison with "natural history" of a condition, or with selected "historical controls". There are, indeed, situations where the patient himself or the natural history of the condition is an appropriate control.
Examples include rare diseases with a high early mortality, and -
particularly for historical controls - conditions like iron for iron
deficiency anemia, liver or vitamin B12 for pernicious anemia, peni-
cillin for bacterial endocarditis, insulin for diabetes mellitus, and
so forth. When one examines reports of trials with "historical controls" it becomes apparent that they are regularly biased against the
control series. Even when they reach the same directional result as
an appropriately designed study, the trial treatment tends to be better
than the "historical control" treatment to an extent far outstripping
the differences in appropriately controlled trials of the same treatment. The presence of bias - conscious or unconscious - is apparent.

Apart from the difficulty of resolving the question of conscious bias in selecting "historical controls", in chronic disease there is always an element of unconscious time bias that tends to invalidate historical controls. Time bias involves three factors: (1) changing disease incidences, for example, coronary artery disease; (2) changing disease severities or patterns, for example, acute rheumatic fever; (3) changing ancillary management, for example, acute myocardial infarction - reduction of bed rest, along with constantly improving treatments for arrhythmias and myocardial failure.

If one were to examine secular trends in coronary heart disease incidence, one would see for the United States, for example, that since 1968 coronary disease has steadily declined in all population subgroups. (Moreover, comparisons of necropsy data in combat casualties among the US troops in the Vietnam War as compared with the Korean War 20 years later showed a striking reduction in the coronary lesions in these young men). If one then examines the death rate from coronary heart disease in the United States between 1968 and 1977 as compared with the expected rate projected from the trend up to 1968, one would have found a decrease in observed deaths - 191,500 fewer deaths than predicted. During this period, therefore, had one introduced any new treatment for coronary disease (even acqua pura) without a control series, the sharply reduced death rate could be claimed as due to that treatment.

Irrelevance of Physicians' Training

It is important to emphasize that the level of intellect and skill of physicians applying a new trial treatment is virtually irrelevant. Most ultimately discarded treatments were introduced by physicians of the highest intellect and skill who produced faulty investigative designs. An example of the irrelevance of skill and intellect is gastric freezing for peptic ulcer disease. This was introduced by a distinguished surgeon and presented in a distinguished forum, the Moynihan lecture of the Royal College of Surgeons. There was apparent success in an uncontrolled trial in 841 patients in 10 hospitals. (2,500 gastric freezing machines were then sold). However, 160 patients in five hospitals randomized to gastric freezing (82 patients) versus a sham procedure using tap water (78 patients) showed no significant differences at $1\frac{1}{2}$, 3, 6, 12, 18 and 24 months.

THE BEHAVIORAL CASE

Associated with the scientific basis for the randomized controlled trial is the behavioral basis. By insisting on the scientific inescapability of controlled trials, the physicians' behavior is made to conform to the best scientific standards. This is because of the long history of behavioral lapses in this area, falling under three general circumstances: (1) <u>Prescription despite lack of efficacy</u>. Examples include stilbesterol (DES) for habitual abortion - controlled trials showed that it was useless; diet in acute peptic ulcer - the ordinary hospital diet caused a higher pH than did the standard "white" ulcer diet; bed rest in acute viral hepatitis - shown to have no effect on prognosis in patients who felt well enough to be ambulatory; the general notion that "it can't do any harm" - e.g. the unthinking use of corticosteroids. (2) <u>Prejudging efficacy</u>. Examples include: diversion of patients from and biasing patient assignment to controlled trials, as well as counseling non-participation. (3) <u>Nonconcern with methodology</u>. At the World Congress of Gastroenterology, 3,000 physicians attended the session on which operation was best and only 300 attended the session on how to determine the best operation.

Unacceptable behavior by physicians is also shown in statistics for mortality from bleeding peptic ulcer between 1930 and 1959. In each of those decades the mortality remained unchanged - approximately 10%. Yet in each of those decades the proportion of patients given surgical treatment steadily increased despite no effect on that mortality.

THE ETHICAL CASE

The ethical case for controlled trials rests on several bases. One of these is that the scientifically optimal method is ipso facto the ethically optimal method. But at an even more fundamental level, on "Day One" of the application of any promising treatment in human beings the investigator - surgeon or nonsurgeon - does not know with precision what will happen in the short run and hasn't the slightest inkling of what would happen in the long run. (With DES patients the real problem occurred in the next generation). Therefore, since the physician is always ethically bound to provide the best available treatment, and cannot know this about new treatments, patients in a trial should be completely informed and have an equal opportunity to get the best treatment. This means an equal opportunity to get either the best standard treatment or the treatment under investigation. To do this in a scientifically and ethically valid manner, randomization from the very first patient is required. Finally, because the randomized controlled trial is the most powerful means of obtaining a true answer, it is unethical not to do a controlled clinical trial when one is feasible.

In summary, coronary bypass surgery is a unique form of treatment but it remains a treatment and therefore subject to the scientific and ethical rules for all treatments. Scientific, behavioral and ethical considerations make mandatory appropriately designed controlled trials no less for coronary bypass surgery than for other treatments.

REFERENCES

Chalmers T.C., 1974, The impact of controlled clinical trials on
        the practice of medicine. Mt. Sinai J. Med., 11:753-759.
Spodick D. H., The randomized controlled clinical trial: Scientific
        and ethical bases. Am. J. Med. (In press).
Spodick D. H., 1982, Randomized controlled clinical trials: The
        behavioral case. JAMA ; 247:2258-2260.
Spodick D. H., 1971, Revascularization of the heart - Numerators in
        search of denominators. (Editorial) Am. Heart J.; 81:149-157.
Spodick D. H., 1973, The surgical mystique and the double standard:
        Controlled trials of medical and surgical therapy for cardiac
        disease: Analysis, hypothesis, proposal. Am. Heart J.; 85:
        579-583.
Spodick D. H. , 1977, Aortocoronary bypass surgery: Emerging triumph
        of controlled clinical trials. Chest; 71:318-319.
Sacks H., Chalmers T. C. and Smith H., 1982, Randomized versus his-
        torical controls for clinical trials. Am. J. Med.; 72:233-240.

SURGERY FOR MYOCARDIAL ISCHEMIA AND

COMPLICATIONS OF MYOCARDIAL INFARCTION

Denton A. Cooley

Texas Heart Institute of St. Luke's Episcopal and
Texas Children's Hospitals
Houston, Texas

The era of open-heart surgery by extracorporeal circulation began during the 1950's[1-4]. As techniques of bypass became simplified and as non-blood prime techniques were introduced, the method came into widespread use. Not until the late 1970's, however, when coronary bypass surgery was introduced, did expansion in open-heart surgery extend to hospitals which previously had not been identified as centers for treatment of heart disease. In the year 1980, 171,667 open-heart operations were performed in 636 hospitals throughout the United States, and fully two-thirds were for treatment of coronary arterial disease and its complications[5].

At the Texas Heart Institute during 1980 and 1981, approximately 5,000 open-heart operations were performed annually (Table 1). Although the numbers of operations for congenital anomalies, valvular heart disease, etc. have remained more or less constant, the number of coronary bypass procedures accounted for the bulk of the expansion (Figure 1). During 1981, for example, among the 4,916 patients undergoing open-heart surgery, 3,568 had coronary artery bypass. During that same year, 182 patients underwent percutaneous transluminal coronary angioplasty. Among these, 10 patients required subsequent coronary bypass.

TECHNIQUE

The emphasis during the 25 years of experience at the Texas Heart Institute has been upon simplification[6,7]. Thus, we have introduced disposable plastic oxygenators, the use of glucose-Ringer's lactate solution for blood prime, and emphasis upon expeditious and efficient technique. The recent introduction of cold

179

Table 1.   Open-Heart Operations

1981	
TOTAL	4,916
Adult	4,663
Pediatric	253
Coronary Bypass	3,568
PTCA	182

Abbreviations: PTCA = percutaneous transluminal
coronary angioplasty.

cardioplegic solutions containing potassium chloride has improved
results.   Use of 500cc of balanced electrolyte solution containing
20mEq of potassium chloride injected into the coronary circulation
at $4^{\circ}C$ produces a quiet, bloodless operative field essential to
precise technique (Figure 2).   Moreover, the standard use by the
surgeon of optical loupes with 3-4 power magnification and improved
illumination enhanced by fiberoptic lights has further improved
surgical technique.

The objective of current techniques for revascularization is
toward reestablishment of a pulsatile flow in the peripheral branches
of the coronary artery.   Thus, formerly indirect methods aimed at
increasing collateral circulation have been entirely discarded.
Preoperative assessment depends largely upon selective arteriography
as introduced by Mason Sones some 15 years ago[8].

Our choice for the bypass conduit is the long saphenous vein.
During operation, the vein is removed using multiple incisions along
the medial aspect of the leg and thigh[9].   The vein is removed, the
tributaries are ligated, and the vein is distended with autologous
heparinized blood.   When the saphenous vein is not present, brachial
veins are the next choice.   The internal mammary artery provides
advantages in some patients and is used as a pedicle to the left
anterior descending coronary artery (Figure 3).   The principal advantage of this vessel is that it maintains patency better than the
free graft of autologous vein.

PATIENT SELECTION FOR REVASCULARIZATION

Selection of patients for coronary bypass is based upon evidence
of at least 50% occlusion of a coronary artery[10].   Usually single
vessel disease must reveal at least a 90% obstruction before operation

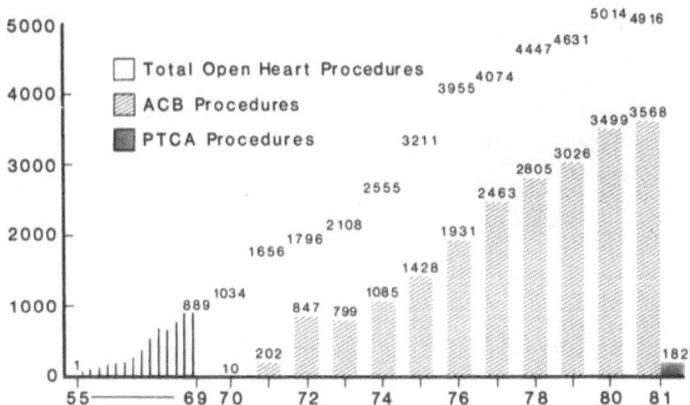

Fig. 1. Number of patients undergoing cardiac surgery at the Texas
         Heart Institute since 1955. The number of operations for
         congenital anomalies and acquired valvular lesions has
         remained more or less stable with the number of aorto-
         coronary bypass procedures accounting for the expansion.

is indicated. We believe that if a patient is selected for cardiopulmonary bypass, a complete revascularization should be done. In
the early experience, end-to-side distal anastomoses were used
routinely. This proved to be technically difficult, since many
patients required 4, 5 or 6 bypass grafts. A solution to this
problem has come with the development of a sequential technique, in
which side-to-side anastomoses are effected to smaller branches of
either the circumflex, left anterior descending, or right coronary
artery.

     Coronary endarterectomy is used primarily for opening the right
coronary artery. When necessary, endarterectomy is also done on
the anterior descending or a marginal branch of the circumflex, but
the incidence of subsequent thrombosis of those vessels may produce
catastrophic results. Right coronary endarterectomy, however, is
usually associated with long-term patency, and provides the surgeon
with an additional opportunity to revascularize a major coronary
artery which would not otherwise be done.

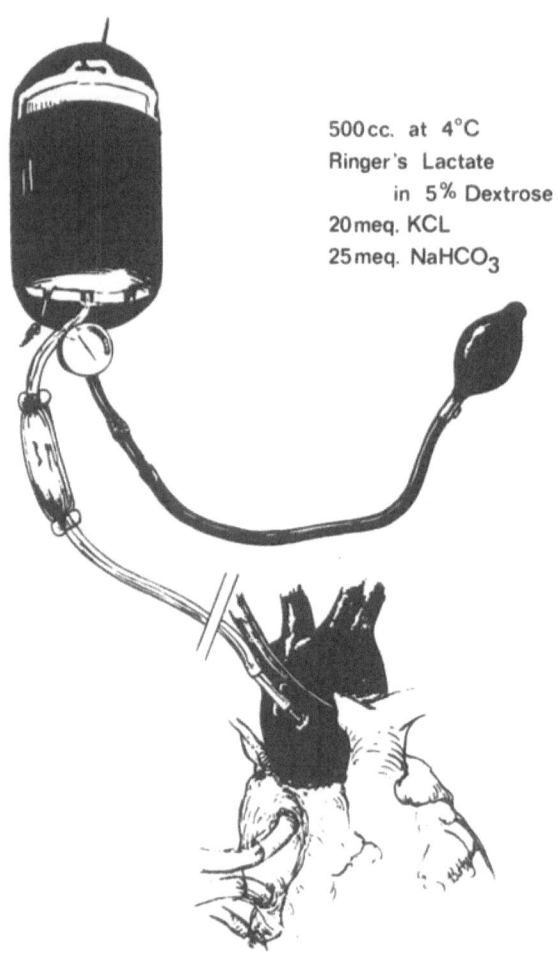

500 cc. at 4°C
Ringer's Lactate
    in 5% Dextrose
20 meq. KCL
25 meq. NaHCO$_3$

Fig. 2.   Injection of cold cardioplegic solution into the ascending
         aorta following distal cross-clamping. This technique
         produces a quiet, bloodless operative field.

COMPLICATIONS OF MYOCARDIAL INFARCTION

     Myocardial infarction producing necrosis of myocardial tissue
may result in serious damage to the ventricle. While many of these
complications can be avoided by proper early treatment of the

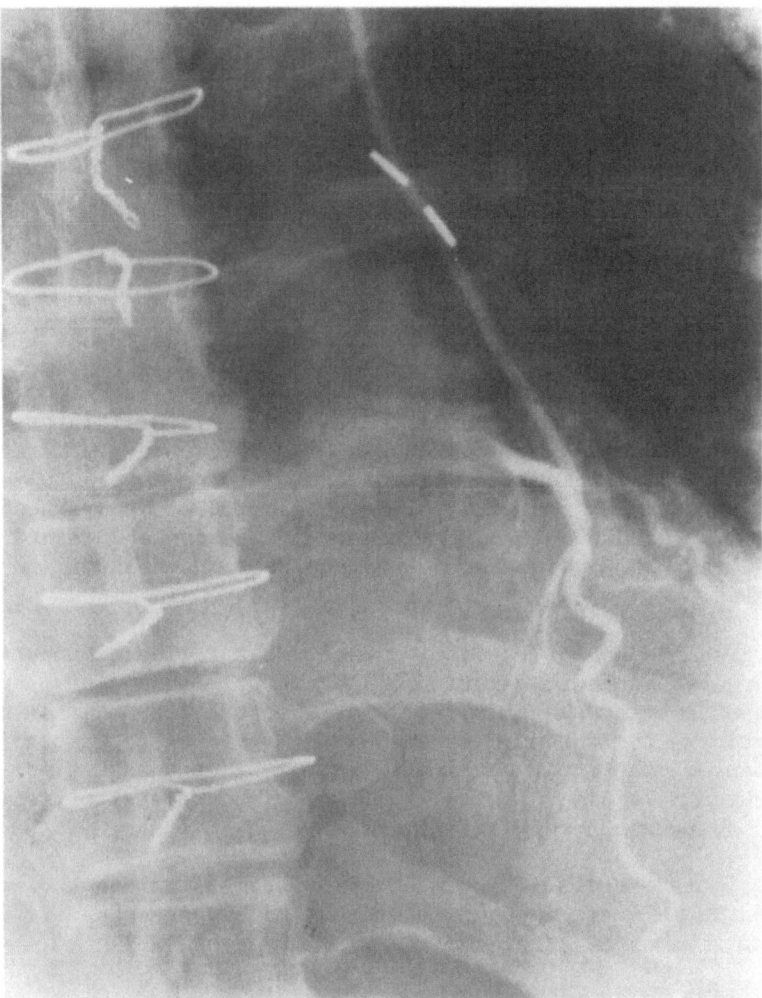

Fig. 3.   Cineangiogram showing internal mammary artery used as a
          pedicle to the left anterior descending coronary artery.

infarction, in the more extensive cases the necrosis may lead to
serious ventricular debility[11].

Acute Cardiac Rupture

Acute cardiac rupture is a complication of myocardial infarction
which usually is not amenable to surgery, although it is possible
that in an occasional case, a repair may be effected. In some instances, fortunately, a pseudoaneurysm occurs; this usually happens
when the patient had a preexisting pericarditis which produced ad-
hesions between parietal and visceral pericardium. In some instances,
the two-staged heart transplant would be a possible alternative,
using a total artificial heart as the first stage with a delayed
cardiac transplantation as the second[12].

Myocardial Aneurysm

One of the commonest lesions following myocardial infarction is
myocardial aneurysm (Figure 4). Although the extent of the lesion
may indicate that subsequent rupture could occur, this rarely happens.
The threat to the patient, therefore, is not from rupture, but from
left ventricular dysfunction, pulmonary congestion and cardiac decompensation. Some patients develop peripheral embolism from intra-
ventricular thrombi.

Surgical repair of the lesion requires total cardiopulmonary
bypass and cardioplegic solution. The aneurysm should be opened to
the left of the interventricular groove, if the lesion is anterior
or posterior. The thrombus is evacuated. If there is extensive
involvement of the interventricular septum, an internal plication
is done with multiple sutures (Figure 5). The lateral wall is then
plicated over Dacron felt strips.

The intraaortic balloon counterpulsation technique has proved
particularly valuable when needed in the early postoperative course
following removal of extensive lesions[13].

Mitral Regurgitation

Myocardial infarction can produce mitral regurgitation either
by causing dilatation of the mitral annulus, ischemia of the papillary
muscles producing irregular closure of the mitral leaflets or frank
rupture of a papillary muscle or chordae tendineae. Each case must
be evaluated according to the findings. For the dilated annulus or

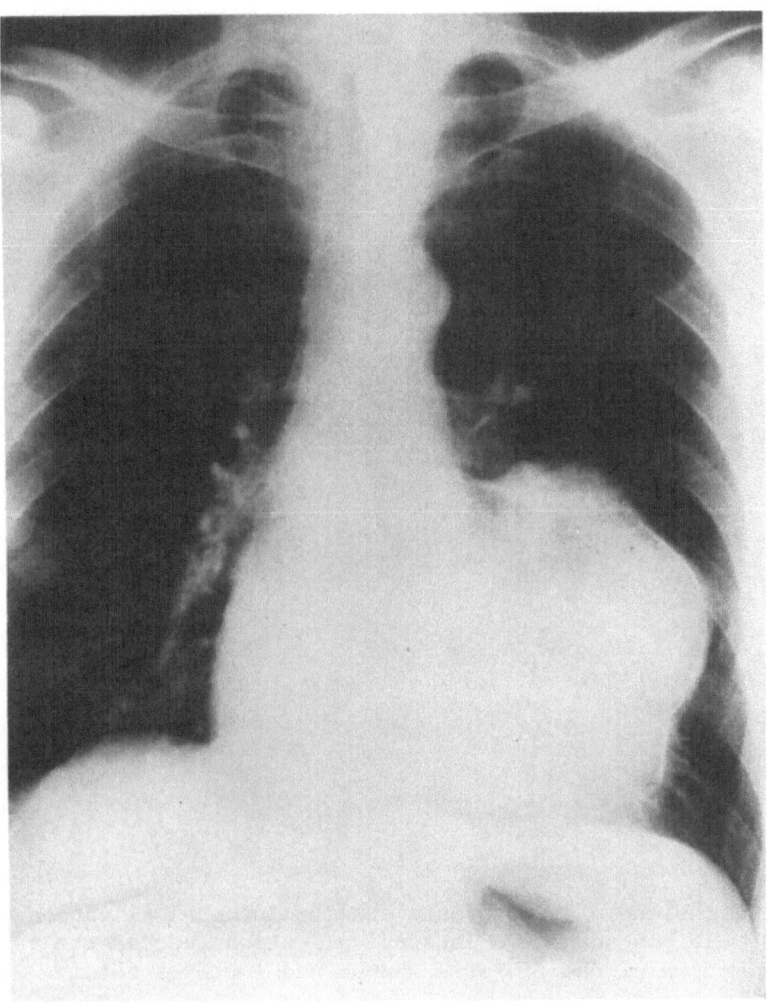

Fig. 4. Radiograph showing ventricular aneurysm prior to surgical resection.

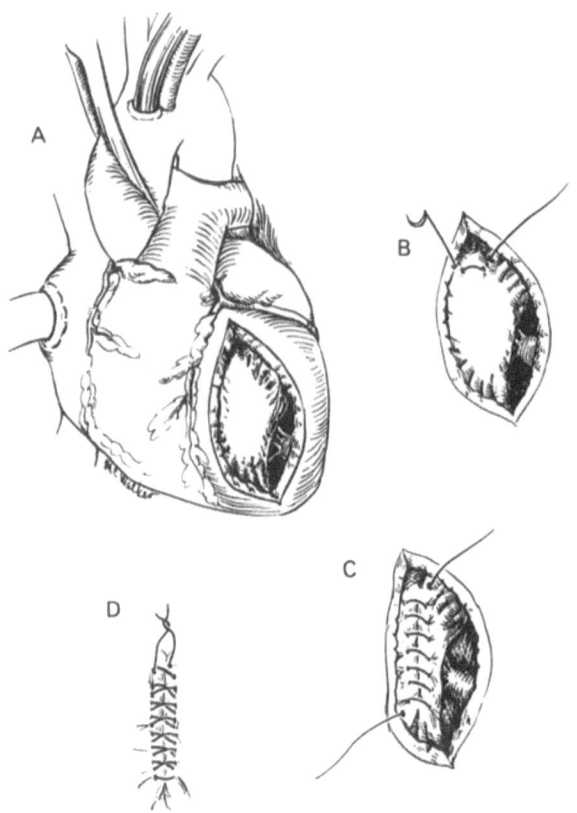

Fig. 5.   Technique of septoplasty showing internal plication with
          multiple sutures. Internal plication reduces the paradoxical
          motion of the fibrotic septum and improves cardiac output.

a ruptured chorda on the posterior leaflet, we prefer an annuloplasty technique. For instances in which a major papillary rupture has occurred or when the anterior leaflet is flail, mitral valve replacement is recommended. At present, we prefer the pericardial xenograft prosthesis of Ionescu-Shiley.

Ventricular Septal Defect

Perforation of the septum during acute myocardial infarction can be catastrophic. Most patients die within the first 48 hours. However, some will survive this period and go on to a chronic state.

Our experience indicates that if surgery can be delayed for four weeks or more, then the possibility of a successful surgical repair is greatly increased. Unfortunately, many patients are in irreversible congestive heart failure at the time when surgical consultation is requested. Surgery, therefore, must then be done on an emergency basis.

In these cases, a ventriculotomy is made over the site of the myocardial infarction, usually, therefore, in the infarcted left ventricle. The ventricular septal defect is examined and repaired by whatever means seems appropriate, generally with interrupted mattress sutures reinforced by an overlying Dacron fabric patch. Most patients with extensive acute myocardial infarction will require counterpulsation support during the early postoperative period.

Dysrhythmia

Paroxysmal ventricular tachycardia and ventricular fibrillation are frequent complications of acute myocardial infarction. Approximately 50% of these patients can have their dysrhythmia corrected by myocardial revascularization. Removal of a ventricular aneurysm may also reverse the tendency to dysrhythmia.

With the current interest in electrophysiology, electrical mapping of the epicardium and endocardium may be performed to identify the focus causing the dysrhythmia. Excision of the internal focus is then recommended. From a practical standpoint, we have found that an encircling endocardiotomy is the technique of choice, making an incision into the endocardium approximately 4-5mm deep (Figure 6). The incision is made proximal to visible endocardial fibrosis or ventricular aneurysm.

Other techniques may be useful in correcting dysrhythmias, including electronic pacemakers to control tachycardia. Recently, interest has been shown in an implantable electronic defibrillator[14].

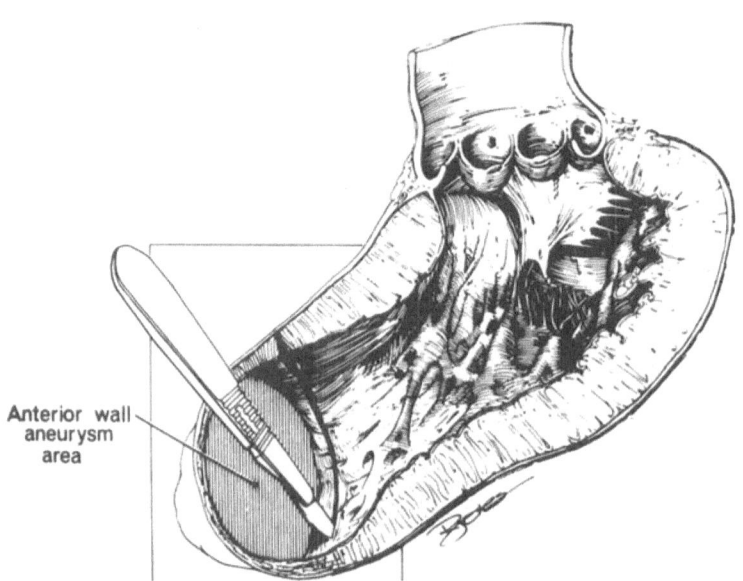

Anterior wall
aneurysm
area

Fig. 6.   Drawing showing the technique of encircling endocardial
          myotomy used to interrupt the retrograde or reentry passage
          of electrical impulses from the scarred myocardium.

COMMENT

     Current techniques of myocardial revascularization and treatment
of complications of myocardial infarction have provided an almost
unlimited source of patients for open-heart surgery. At the Texas
Heart Institute since 1969, a total of 26,404 patients have undergone open-heart surgery associated with the coronary bypass procedure.
The overall hospital mortality in this group has been 4.2%. Among
21,821 patients having only coronary bypass performed without concomitant procedures, the hospital mortality was 3.0%.

     The presence of additional lesions such as a ventricular
aneurysm, ventricular septal defect or mitral regurgitation increases
the risk of surgery, so that the risk may be 8-10%, depending on the
extent of the lesion. Nevertheless, the treatment of coronary heart
disease has been greatly enhanced by the direct surgical approach.

REFERENCES

1. C.W. Lillehei, M. Cohen, and H.E. Warden, et al., The direct-vision
    intracardiac correction of congenital anomalies by controlled
    cross circulation, Surgery 38:11 (1955).
2. D.A. Cooley and M.E. DeBakey, Resection of the entire ascending
    aorta in fusiform aneurysms using cardiac bypass, JAMA 162:1158
    (1956).
3. D.A. Cooley, B.A. Belmonte, M.E. DeBakey, and J.R. Latson,
    Temporary extracorporeal circulation in the surgical treatment
    of cardiac and aortic disease: Report of 98 cases, Ann.Surg.
    145:898 (1957).
4. R.D. Bloodwell, J.N. Kidd, and G.L. Hallman, Valve replacement
    without coronary perfusion: Clinical and laboratory observations,
    in: "Prosthetic Heart Valves," L.A. Brewer ed., Springfield Ill
    Thomas, 1959, pp. 397-418 (1957).
5. National Heart, Lung and Blood Institute's Fact Book, US Dept
    of HEW, National Institutes of Health, NIH Publication No.80-
    1674, Bethesda, Maryland. (1980).
6. D.A. Cooley, Revascularization of the ischemic myocardium,
    J.Thorac.Cardiovasc.Surg. 78(2):301 (1979).
7. F.M. Sandiford, D.A. Cooley, and D.C. Wukasch, The aortocoronary
    bypass operation: An overview based on 10,000 operations at the
    Texas Heart Institute, Intl.Surg. 63(4):83 (1978).
8. F.M. Sones Jr., and E.K. Shirey, Cinecoronary arteriography.
    Mod.Concepts Cardiovasc.Dis. 31:735 (1962).
9. D.A. Cooley and J.C. Norman, in: "Techniques in Cardiac Surgery,"
    pp. 153-156 Texas Medical Press, Houston (1975).
10. D.C. Wukasch, D.A. Cooley, R.J. Hall, G.J. Reul, F.M. Sandiford,
    and S.L. Zillgitt, Surgical versus medical treatment of coronary
    artery disease: Nine-year follow-up of 9,061 patients, Am.J.
    Surg. 137:201 (1979).
11. G.J. Reul, D.A. Cooley, F.M. Sandiford, E.R. Kyger, D.C. Wakasch,
    and G.L. Hallman, Aortocoronary artery bypass: Present indi-
    cations and risk factors, Arch.Surg. 111:414 (1976).
12. D.A. Cooley, Staged cardiac transplantation: Report of three
    cases, Heart Transplantation 1(2):145 (1982).
13. S.R. Igo, C.W. Hibbs, R. Trono, J.M. Fuqua, C.H. Edmonds, C.J.
    Leachman, M.A. Brewer, D.A. Holub, and J.C. Norman, Intraaortic
    balloon pumping: Theory and practice. Experience with 325
    patients, Artificial Organs 2(3):249 (1978).
14. L. Watkins Jr., M. Mirowski, M.M. Mower, P.R. Reid, L.S.C.
    Griffith, S.C. Vlay, M.L. Weisfeldt, and V.L. Gott, Automatic
    defibrillation in man: The initial surgical experience, J.Thorac.
    Cardiovasc.Surg. 82(4):492 (1981).

SURGERY FOR MYOCARDIAL ISCHEMIA: PRESENT INDICATIONS, TECHNICAL ASPECTS AND RESULTS. AN ANALYSIS OF 26,404 PATIENTS

Denton A. Cooley

Division of Surgery, Texas Heart Institute of
St. Luke's Episcopal and Texas Children's Hospital
Houston, Texas

ABSTRACT

Between October 1969 and December 31, 1981, 26,404 patients
with coronary insufficiency underwent direct myocardial revascularization using aortocoronary bypass (ACB). Among these patients,
21,821 had ACB alone, and the remaining 4,583 had ACB in addition
to correction of other cardiac and vascular lesions. In the series
of patients having ACB alone, the hospital (early) mortality was
3.0%, with a 2.1% early mortality for 1981. Total (early) mortality
was 4.2%, with a 3.9% early mortality for 1981. Total hospital
mortality was higher for women (7.3%) than for men (3.9%), but late
survival was approximately the same for both sexes.

Among surviving patients, 93% were improved or asymptomatic
after undergoing ACB. According to our data, at the end of 10 years
95% of the patients will be survivors. These results indicate the
benefit of surgical treatment for both symptomatic relief and increased life expectancy.

Coronary artery occlusive disease and myocardial ischemia are
the leading causes of death in Western societies. Aortocoronary
artery bypass (ACB) has now been accepted by most cardiologists as
appropriate treatment for coronary artery occlusive disease (CAOD).
The major indication for ACB has been the relief of angina pectoris.
Risk factors have been determined by chronicity, severity of the
underlying coronary artery occlusive process, amount of myocardial
destruction, and presence of other problems, e.g., diabetes, hypertension, etc. Stages in the evolution of coronary artery surgery
have necessarily altered these surgical indications and risk factors.
Since enough patients are now available not only for determination

of risk factors, but also for evaluation of actuarial survival, more
objective evidence has been accumulated. The purpose of this report
is to describe the status of ACB and to review the present indications, surgical techniques and results of our experience with myocardial revascularization since 1969.

INDICATIONS FOR CORONARY ARTERY BYPASS SURGERY

The primary indication for myocardial revascularization in our
series was disabling angina pectoris, although consideration was
also given to patients with cardiac failure resulting from ventricular
aneurysm, ventricular septal perforation, papillary muscle rupture,
mitral incompetence and uncontrolled dysrhythmia. The indications
for operation were mostly the presence of angina pectoris, adequate
distal coronary arteries, as shown angiographically, viable myocardium
distal to the coronary obstruction, and adequate left ventricular
function. In our early experience, evidence of a severely compromised
left ventricle, as shown by a raised end-diastolic pressure and reduced ejection fraction, was considered to be a contraindication
for operation. Poor left ventricular function when accompanied by
angina, however, is no longer considered an absolute contraindication
to bypass, unless there is angiographic evidence of diffuse disease
in the distal coronary arteries, which would preclude a technically
satisfactory anastomosis.

A second group of patients in whom operation is indicated are
asymptomatic patients with ischemia on treadmill exercise testing,
but with angiographic demonstration of proximal occlusive lesions
in one or more major vessels, particularly the left main coronary
artery. Obstruction of blood flow by an arteriosclerotic plaque depends not only on the degree of stenosis produced, but also on the
length of the lesion.

We believe that preinfarction angina constitutes another indication for emergency ACB. Patients with uncontrollable ventricular
dysrhythmias associated with myocardial ischemia also may benefit
from emergency revascularization. Patients experiencing intractible
cardiogenic shock after acute myocardial infarction can often be
saved by inserting an intraaortic balloon pump, followed by immediate
coronary arteriography and emergency revascularization[1].

Patient Selection

The major determinants of patient selection are the clinical
presentation of the patient, the coronary artery anatomy as demonstrated by coronary arteriogram and the presence of associated disease.

Symptoms of congestive heart failure may be relieved by coronary
artery bypass if myocardial ischemia is the cause of the congestive

heart failure. In some cases, it is reversible. In most cases of
longstanding disease with multiple myocardial infarctions and global
myocardial dyskinesia, congestive heart failure will not be relieved.
In patients with complications of coronary artery disease such as
ventricular aneurysm, ventricular septal defect or mitral insufficiency, combined surgery may result in relief of congestive heart
failure.

Mitral insufficiency secondary to papillary muscle necrosis or
dysfunction usually results from a massive posterior and lateral wall
infarction. Most of these patients have ischemic myocardiopathy and
multivessel disease with previous infarctions and are one of the
highest risk groups who undergo surgery. Valve replacement or mitral
annuloplasty with concomitant coronary artery bypass surgery is recommended for these patients.

Left ventricular aneurysm resection is combined with coronary
artery bypass or is done alone in patients who are symptomatic
following myocardial infarction. In those patients with severe
anatomical disease in other vessels and clots in the left ventricle,
coronary artery bypass surgery may be indicated with prophylactic
excision of the left ventricular aneurysm and removal of the clots.

Acute ventricular septal defect may occur following a septal
infarction. The patient may have acute pulmonary edema and require
immediate surgical intervention. If the cardiac function can be
stabilized, it may be advantageous to wait for surgery for one or
two weeks from the time of the acute phase.

One of the primary determinants in patient selection for surgery
is the anatomical nature of the coronary artery disease, as determined
by coronary arteriogram. An anatomically significant obstruction is
present when 75 percent or more of the lumen is encroached upon by
plaque in a localized or diffuse area. Long areas of obstruction
are probably more significant than short areas. We believe that all
lesions producing approximately 50% stenosis in other arteries
should be bypassed at the time of ACB, because in our experience,
non-bypassed 50% lesions may progress to higher stenoses and are
one of the major factors in patients requiring reoperation following ACB due to progression of disease[2,3]. There has been no evidence
in our series that a bypass to a noncritically occluded artery will
develop thrombosis due to competitive flow. Graft failure and the
progression of atherosclerosis to a critical lesion were the most
common reasons for reoperation. Long-term follow-up showed a peak
of graft failure and progression of disease in nongrafted arteries
at three years and decreased thereafter. Progression of distal
disease in grafted arteries was not temporally related and was uncommon.

When there is total occlusion, it is important to estimate the
collateral circulation and the distal runoff. In most instances,

these are much smaller than they are at the time of surgery because
of poor collateral flow. It is also essential to estimate the
nature of the occlusive disease, i.e., whether it is diffuse and
distributed throughout the coronary arteries extending into the distal
arteries and whether or not the distal circulation will support a
bypass.

Significant stenosis of the left main coronary artery is an
almost unqualified indication for surgery. Occlusion of the right
coronary artery with significant left main stenosis may result in a
true surgical emergency in the patient with unstable angina. Most
occluded right coronary arteries can be bypassed regardless of the
distal circulation, since endarterectomy may be accomplished on this
artery, thus opening the entire distal circulation. The left anterior
descending is probably the most important artery controlling the
anterior septal portion of the left ventricle and is most frequently
involved in disease. One or more diagonal branches may also be important for perfusion of the lateral wall. The main circumflex and
its three or four branches, including the obtuse marginal branches
and the circumflex posterior descending, are all essential to the
posterior and lateral circulation of the left ventricle.

With the recent success of percutaneous transluminal coronary
angioplasty (PTCA) of the coronary arteries, single vessel disease
may now be treated noninvasively. Failure of balloon dilatation in
these patients usually necessitates immediate surgery. Single vessel
bypass would be indicated in the left anterior descending system if
the lesions were very high grade (95%) above the first septal per-
forator or diagonal branch. It might be considered in the right
coronary artery system if arrhythmias and preinfarction angina were
present.

The only anatomical contraindication to surgery is, perhaps,
the presence of very diffuse arterial or myocardial disease, making
bypass unbeneficial.

SURGICAL TECHNIQUE

The operations were conducted by experienced cardiovascular
surgeons of the surgical staff at the Texas Heart Institute (Drs.
George J. Reul, O. Howard Frazier, Grady L. Hallman, David A. Ott,
J. Michael Duncan, William E. Walker, James J. Livesay, and others).

The methods used in myocardial revascularization at the Texas
Heart Institute have been described in detail previously[4,5]. The
saphenous vein is carefully removed and all branches are ligated in
a manner that does not compromise the main lumen. Distension of the
vein is accomplished with autologous heparinized blood, since the
vein wall may be injured when a saline solution is used. When the

patient has no suitable saphenous veins, the alternative sources of
grafts are brachial veins. Internal mammary arteries are seldom
used in this institution. Vascular substitutes have rarely been
used.

Most operations are performed utilizing cold ischemic cardiac
arrest under temporary cardiopulmonary bypass with hemodilution
techniques. When a localized stenosis of the right coronary artery
is the only lesion present, the anastomoses are performed without
cardiopulmonary bypass. Since January 1977, a cold (5°C) cardioplegic solution consisting of 500cc 5% dextrose and 0.45% sodium
chloride containing potassium chloride 20mM, magnesium chloride
7.5mM, sodium bicarbonate 2.5mM, and calcium chloride 1.0mM has
been injected into the ascending aorta after distal cross-clamping.
After cardiopulmonary bypass is begun by the conventional technique
and the ascending aorta cross-clamped, the heart is arrested and
cooled topically. During the period of cardiac arrest, the heart
is maintained at 20 to 25°C in a non-working collapsed state. Distal
anastomoses are performed with 6-0 monofilament polypropylene continuous sutures by reversing the saphenous vein and attaching it to
the distal coronary arteries beyond the last area of occlusion.
Proximal anastomoses are performed with 5-0 sutures after the aortic
cross-clamp has been released and a partial occlusion clamp has
been applied, permitting cardiopulmonary bypass to continue and
restoring coronary perfusion (Figure 1). The surgeon wears optical
magnifying loupes (3-4 power) and a headlamp with a fiberoptic light
to provide precision in anastomosing small arteries.

The technique of using sequential grafts (one vein anastomosed
side-to-side to two or more arteries) has enhanced the ability of
the surgeon to revascularize all significant lesions (Figure 2).
This technique is particularly applicable for multiple lesions in
adjacent arteries and in patients having a limited length of available
vein and a short ascending aorta, which would make placement of five
or six proximal anastomoses technically difficult. In many of the
patients, the diameter of the vein may be considerably larger than
the diameter of the recipient artery. Therefore, the side-to-side
technique makes it possible to make an anastomosis which is more
appropriate than end-to-side. Endarterectomy is avoided whenever
possible on the left coronary artery, but is often necessary in the
occluded right coronary artery.

COMPLICATIONS

Serious nonfatal complications may occur in about 15 percent
of patients undergoing coronary artery bypass and may result in
death in less than 2 percent[6,7]. Most nonfatal complications are
usually related to preexisting conditions, which can be diagnosed
preoperatively and controlled postoperatively.

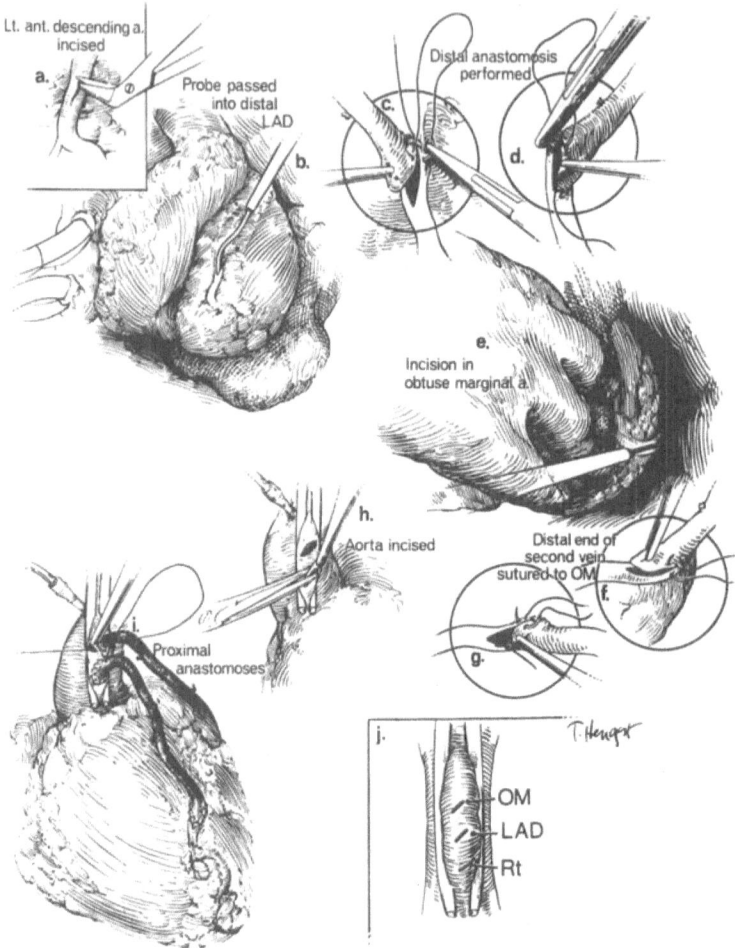

Fig. 1.   The technique for coronary artery bypass to the left
          anterior descending and obtuse marginal branches is
          shown. The inset demonstrates the proximal anastomosis
          to the ascending aorta.

     The severity of presence of diabetes mellitus does not contra-
indicate coronary artery bypass surgery, although there may be an
increase in wound problems, especially if the patient is obese.
Hypertension should be controlled during surgery. Renal hypertension,
which is seen in a few patients, can be corrected after surgery,
particularly if the ischemic syndrome is unstable. Hemorrhagic
disorders which are diagnosed preoperatively can usually be controlled postoperatively by correction of the hemorrhagic state
following surgery. Renal failure in patients without previous renal
disease or low cardiac output rarely, if ever, occurs.

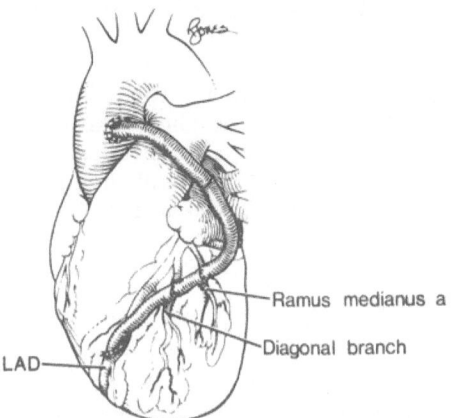

Fig. 2. The technique for sequential anastomosis to three coronary
arteries using one vein graft is shown. Because of the
larger distal flow, there is a higher velocity of flow
through the vein graft.

Atrial fibrillation is the most common arrhythmia seen in the
postoperative period; this is usually controlled by medication.
Heart block may occur, especially in patients with severely diseased
right coronary arteries without run-off. In these patients, temporary pacing wires are placed and the heart block is controlled by
cardiac pacing. An A-V sequential pacemaker may be utilized for
patients with low cardiac output who require pacing and "atrial kick"
for cardiac output.

Nonfatal perioperative infarction occurs in less than 2% of
patients after routine coronary bypass. In most of these patients,
there is no evident clinical difference; diagnosis is usually made
by electrocardiogram, enzyme elevation and left ventricular performance studies[8].

Postoperative hemorrhage requiring return to the operating
room for hemostasis occurs in less than one percent of our patients.

If hemorrhagic diathesis is present, blood components can be used to correct the distorted coagulation. We have operated upon over 700 Jehovah's Witnesses, who do not allow giving blood or blood components during surgery, and have not experienced a significant increase in mortality[9].

Wound infection varies according to the condition of the patient. A higher incidence occurs in women, in the elderly, obese, diabetic and the chronically ill. Wound infection occurs more frequently in the leg. These infections, however, are subcutaneous and can be cured with local treatment. Sternal wound infection, which is more serious, has occurred in less than one percent. Most complicated cases can be corrected with sternal debridement and rewiring followed by long-term antibiotics[10].

Serious neurological complications which result in permanent disability have occurred in less than one percent. The recent decrease in complications has occurred as a result of improved cardiopulmonary bypass techniques and proper postoperative evaluation with appropriate treatment of carotid artery stenosis. Since air is not introduced into the heart during surgery, air embolus is not a problem. In patients with severe arteriosclerosis, postoperative cerebral embolus may occur just as following any other surgical procedure.

CLINICAL DATA

Between October 1969 and December 1981, a consecutive series of 26,404 patients with coronary insufficiency had myocardial revascularization using aortocoronary bypass at the Texas Heart Institute. No patients were excluded from the series because of high risk factors or severity of disease. The series comprised 21,821 patients who had ACB alone, and 4,583 who had concomitant surgical procedures for the treatment of cardiovascular dysfunction (Table 1). In the group with ACB alone, 86% were males and 14% were females. Most patients were between 40 and 60 years of age. Since we believe that every stenotic coronary artery whose structure permits a satisfactory anastomosis should be bypassed, in the entire series 70% underwent three or more bypass grafts (Table 2).

RESULTS

Early Mortality

Early mortality (hospital death) has declined during each year of the study, apparently as a result of increasing technical experience rather than stringent selection. During the year 1970, the first year in which significant numbers of operations were performed,

Table 1. Coronary Bypass Surgery (October 1969 –
          December 31, 1981)

Procedure	No. of Patients
Coronary bypass alone	21,821
Left ventricular aneurysm resection	1,764
Aortic valve procedure	885
Mitral valve procedure	663
Aortic and mitral procedures	70
Other associated procedures	671
Reoperation	530
TOTAL	26,404

early mortality for all patients undergoing ACB alone was 9.7% (17 deaths among 175 patients). During 1981, early mortality in patients undergoing ACB alone declined to 2.1%. Early mortality among patients undergoing associated procedures in 1981 was 15%, and the mortality of the entire series was 3.9% (Table 2).

Women with coronary artery occlusive disease represent a significantly higher risk as shown in Table 3, which shows an early mortality rate of 5.7 in women compared with 2.7 in men. This may be attributable to the presence of smaller arteries and a higher incidence of diabetes and hypertension. The higher early risk in women, however, did not extend late into the follow-up, since the late mortality was similar for both sexes (Table 3).

The actual number of bypasses performed reflects the extent and severity of the coronary disease and degree of myocardial fibrosis and ventricular dyskinesia. Early mortality was higher when three or more vessels were bypassed. These patients were a higher risk group because they had more diffuse disease and required more prolonged surgical procedures. Late mortality, however, occurred less often in those patients in whom more complete revascularization was performed (Table 3). This supports our policy for bypassing all diseased vessels.

Symptomatic Results

A long-term assessment of symptoms up to 10 years after operation was made in 12,276 patients in whom follow-up could be obtained. This showed that 93.2% of patients remained asymptomatic or signifi-

Table 2. Early Mortality related to number of vessels grafted
          and concomitant procedures

	10/69-12/31/81			1981		
	Patients	Early Deaths	%	Patients	Early Deaths	%
Alone	21,821	648	3.0	3,082	66	2.1
Single	1,544	34	2.2	132	--	----
Double	5,080	159	3.1	445	12	2.7
Triple	8,679	306	3.5	1,077	30	2.8
Quadruple	5,220	124	2.4	1,091	19	1.7
Quintuple	1,175	22	1.9	296	5	1.7
Sextuple or more	123	3	2.4	41	--	----
LV aneurysm	1,764	150	8.5	251	29	11.6
Aortic procedure	885	112	12.7	146	11	7.5
Mitral procedure	663	91	13.7	97	14	14.4
Aortic and mitral procedure	70	17	24.3	8	3	37.5
Other procedures	671	59	8.8	123	13	10.6
Reoperation	530	49	9.3	140	12	8.6
TOTAL	26,404	1,126	4.2	3,847	148	3.9

cantly improved, 3.6% were symptomatically unchanged, and 3.2% were
worse (Table 4). Significantly, the symptomatic results improved
when more complete revascularization was performed.

Long-Term Follow-Up

The cumulative survival of the entire group of patients (from
follow-up), calculated by actuarial methods, has been 91% at three
years, 88% at five years and 70% at ten years, yielding an average
attrition rate (including early mortality) of 3.12% per year. The
average attrition rate of the survivors after the first month has
been 2.26% per year. It should be noted that in these patients
mortality is presented cumulatively without regard to cause.

Table 3. Comparison of early and late Mortality related to sex and
          number of vessels grafted (October 1969 to December 31,
          1980)

	Patients	Early Deaths	%	Early Deaths	%
Men					
Single	1,053	23	2.2	55	5.2
Double	3,893	103	2.7	250	6.4
Triple	6,585	209	3.2	257	3.9
Quadruple	3,728	80	2.2	71	1.9
Quintuple	824	16	1.9	10	1.2
Sextuple or more	72	3	4.2	2	2.8
TOTAL	16,155	434	2.7	645	4.0
Women					
Single	359	11	3.1	15	4.2
Double	742	44	5.9	33	4.5
Triple	1,017	67	6.6	39	3.8
Quadruple	401	25	6.2	10	2.5
Sextuple or more	10	--	---	--	---
TOTAL	2,584	148	5.7	99	3.8
TOTAL (Men and Women)	18,739	582	3.1	744	4.0

Approximately 2/3 of the patients died of cardiac causes and the
remainder died of other causes.

Causes of Death

    The most common causes of death in both males and females were
intraoperative or postoperative myocardial infarction, cerebrovascular
accidents, arrhythmias, and congestive heart failure. The causes of

Table 4. Symptomatic results in patients surviving operation by
          follow-up (October 1969 - December 31, 1980).

	No. of Patients	Asymptomatic (%)	Improved (%)	Same (%)	Worse (%)
Alone	10,589	38.6	54.6	3.7	3.1
Single	919	33.5	55.4	5.4	5.7
Double	2,928	33.7	56.6	5.4	4.3
Triple	4,415	40.0	54.7	2.8	2.5
Quadruple	1,929	43.4	52.2	2.6	1.8
Quintuple	364	48.1	48.1	1.1	2.7
Sextuple or more	34	52.9	47.1	---	---
LV Aneurysm	641	30.1	63.2	2.5	4.2
Aortic procedure	404	43.8	50.7	2.5	3.0
Mitral procedure	243	27.2	63.8	2.5	6.5
Aortic and mitral procedure	25	32.0	64.0	4.0	---
Other procedures	243	35.4	58.9	2.5	3.2
Reoperation	131	21.4	61.8	9.2	7.6
TOTAL	12,276	37.9	55.3	3.6	3.2

late deaths were myocardial infarction, cerebrovascular accidents,
and congestive heart failure. Therefore, only approximately 50%
of late deaths were cardiac related[11,12].

DISCUSSION

    The reduction in surgical mortality in a decade of experience
from 9.7% to 2.1% is encouraging and emphasizes the importance of
increased surgical experience and improved techniques, particularly
the use of cold cardioplegia, topical cardiac hypothermia, and the
greater technical precision provided by optical magnification and
high intensity illumination. The early mortality rates have improved, even though our indications for ACB have been broadened to
include those patients who have angina pectoris in the presence of

poor ventricular function (ejection fractions less than 0.2). As a
rule, only those patients with minimal or no angina pectoris who
have angiographic evidence of diffuse coronary arteriosclerosis that
would preclude technically satisfactory grafts are refused operation[13].
When higher risk patients are excluded from the series, early mortality is less than 1%, the mortality considered acceptable for most
elective major surgical procedures.

The higher early mortality in women compared with men is striking. Early mortality has been reduced considerably in male but not
in female patients[14]. A possible explanation for this difference
may be that when women develop coronary insufficiency, the metabolic
derangement is more severe, thereby producing more diffuse lesions.
In addition, it has been our impression that coronary arteries in
women tend to be smaller, more friable and more often intramyocardial
in location than in men, thereby making satisfactory anastomoses
technically more difficult. Another factor to be considered is the
higher incidence of diabetes among women. Interestingly, the higher
early mortality in women did not extend to late mortality, which was
approximately equal for both women and men.

The importance of complete revascularization of all stenotic
coronary arteries is shown by the lower late mortality in those
patients in whom this was achieved. We consider the improved long-
term survival outweighs the risk of the more prolonged surgical procedure required for multiple bypass grafts. Additionally, our experience has shown that complete revascularization of all lesions
producing more than 50% luminal stenosis is a major factor in reducing the need for subsequent revascularization procedures. The
results clearly indicate better long-term symptomatic relief with
more complete revascularization.

SUMMARY

It is estimated that each year in the United States approximately
175,000 open-heart operations will be done. About 75-80 percent of
these operations will be coronary bypass procedures. With the good
results obtained over the past decade, this operation has withstood
the test of time. The simple technique of bypassing an area of
obstruction with a new conduit is the only way that a significantly
greater amount of blood can be brought to an ischemic area of the
myocardium. The actual relief of symptoms and prolongation of life
results only with increased blood flow to the myocardium.

On the basis of our experience and observations, coronary artery
bypass provides for significant improvement in both the quality and
length of life in patients with severe coronary artery occlusive
disease. Coronary artery bypass can now be safely performed with
minimal mortality and increased long-term survival.

REFERENCES

1. J.T. Sturm, T.M. Fuhrman, S.R. Igo, D.A. Holub, M.G. McGee, J.M. Fuqua, and J.C. Norman, Quantitative indices of intra-aortic balloon pump (IABP) dependence during post-infarction cardiogenic shock, Artificial Organs 4(1):8 (1980).

2. G.J. Reul Jr., D.A. Cooley, D.A. Ott, A. Coelho, L. Chapa, and I. Eterovic, Reoperation for recurrent coronary artery disease, Arch.Surg. 114:1269-1275 (1979).

3. D.C. Wukasch, M. Toscano, D.A. Cooley, G.J. Reul, F.M. Sandiford, E.R. Kyger, and G.L. Hallman, Reoperation following direct myocardial revascularization, Circulation 56, Suppl. II, 3-7 (1977).

4. D.A. Cooley and J.C. Norman, in: "Techniques in Cardiac Surgery," Texas Medical Press, Houston (1975).

5. D.A. Cooley, Revascularization of the ischemic myocardium, J. Thorac.Cardiovasc.Surg. 78(2):301-304 (1979).

6. R.J. Hall, E. Garcia, V.S. Mathur, D.A. Cooley, K.D. Gold, and A.G. Gray, Factors influencing early and late survival after aortocoronary bypass: A preliminary report, Cardiovascular Diseases 4(2):120-128 (1977).

7. G.J. Reul, D.A. Cooley, D.C. Wukasch, E.R. Kyger, F.M. Sandiford, G.L. Hallman, and J.C. Norman, Long-term survival following coronary artery bypass: Analysis of 4,522 consecutive patients, Arch.Surg. 110:1419-1424 (1975).

8. J.T. Dawson, E. Garcia, R.J. Hall, and D.A. Cooley, Serum enzyme after coronary artery bypass surgery, Circulation 46, Suppl. II: 144 (1972).

9. D.A. Ott and D.A. Cooley, Cardiovascular surgery in Jehovah's Witnesses: Report of 542 operations without blood transfusion. JAMA 238(12):1256 (1977).

10. D.A. Ott, D.A. Cooley, R.T. Solis, and C.B. Harrison, III, Wound complications after median sternotomy: A study of 61 patients from a consecutive series of 9,279, Cardiovascular Diseases 7(1):104-111 (1980).

11. D.C. Wukasch, R.J. Hall, D.A. Cooley, G.J. Reul, J.M. Oglietti, E.R. Kyger, F.M. Sandiford, and G.L. Hallman, Surgical versus medical treatment of coronary artery disease: Long-term survival, Vasc.Surg. 10(5):300-314 (1976).

12. D.A. Cooley, D.C. Wukasch, F. Bruno, G.J. Reul, F.M. Sandiford, S.L. Zillgitt, and R.J. Hall, Direct myocardial revascularization Experience with 9364 operations, Thorax 33(4):411-417 (1978).

13. D.C. Wukasch, D.A. Cooley, G.J. Reul, R.J. Hall, M. Vucinic, F.M. Sandiford, J.C. Norman, E.R. Kyger, and G.L. Hallman, Surgical treatment of angina pectoris: Current status, Angiology 28:169-180 (1977).

14. D.C. Wukasch, R.J. Hall, D.A. Cooley, G.J. Reul, J.M. Oglietti, E.R. Kyger, F.M. Sandiford, and G.L. Hallman, Surgical versus medical treatment of coronary artery disease: Long-term survival, Vascular Surgery 10:300-314 (1976).

SURGICAL MODIFICATION OF MORTALITY FROM

CORONARY ARTERY DISEASE

Robert D. Leachman[*] and
Dennis V. Cokkinos[**]

[*]Baylor College of Medicine
St. Luke's Episcopal Hospital
Texas Heart Institute, Houston, Texas
[**]The University of Athens, Athens, Greece

SURGERY FOR CORONARY ARTERY DISEASE

Coronary arterial obstruction bypass, performed mainly by inverted saphenous vein grafts and to a lesser extent by internal mammary artery anastamosis has been in existence for about 12 years. In 1979, it was estimated that 93,000 operations were done in the United States (Miller et al., 1979). It would be futile here to try to recapitulate the discussions as to the merit of this technique in each category of patients. We shall try to sum up those points in which general agreement exists.

There is agreement among most authors that surgery increases survival in significant main left coronary artery stenosis (Table 1). Some discordant notes have been reported: Conley et al. (1978) differentiated between patients with 50-70% stenosis and those with 70% stenosis. The latter group had a significantly lower survival at one year (72 versus 91%) and at 3 years (41 versus 66%). Moreover, these authors evaluated a subgroup of similar patients with none or just one of these high risk features: congestive heart failure, anginal pain at rest, cardiomegaly, and resting ST-T wave changes. These low-risk patients had a survival of 97% at one and 74% at three years. However, despite these and some other (DeMots et al., 1975) discordant notes, the concensus among cardiologists and surgeons is that left main artery disease, which very seldom is isolated, should be operated upon.

Another point of agreement is that in patients with single coronary artery disease (excepting the main left coronary artery),

205

Table 1. Medical and Surgical Treatment of Left Main Coronary Artery Disease

| Authors | Number of Pts | | Follow-up Survival % Years | | | | | | | |
| | | | 1 | | 2 | | 3 | | 5 | |
	M	S	M	S	M	S	M	S	M	S
Bruschke et al 1973	37	—							57	
Pichard et al 1975	48	52					60	83		
Lim et al 1975	141	—							49	
Zeft et al 1974	—	56				72.2				
Alford et al 1974	—	86		88.4						
Cohen and Gorlin 1975	17	40					40	89		
DeMots et al 1975	19	28	62	78			55	70	44	70
Takaro et al 1977	53	60			71	93				
Talano et al 1975	32	89			61	82				
Sung et al 1975	11	17	55	89						
Conley et al 1978	163	a.50-70%	91				66			
		b.>70%	72				41			
Campeau et al 1978	114	197					60	83	48	80
Lawrie et al 1978		86								90
Hurst et al 1978		88		85						
Greene et al 1977		55						86		
Sheldon 1977		63								83
Loop et al 1979		300								88
Killen et al 1980		271								88
European Study Group 1980	31	28							64	93

Table 2. Medical and Surgical Treatment of Single Coronary
          Artery Disease except for Left Main

Authors	Number of Patients		Follow-up Survival % 5 years	
	M	S	M	S
Bruschke et al. 1973	202		62*-90**	
Kouchoukos et al. 1977	24	29	96	89
Hammermeister et al. 1977	171	316	94	96
Lawrie et al. 1978		182		93
Texas Heart Institute 1981		466		92

* = with 30-50% stenosis of other arteries + occlusion of 1
** = with all other arteries normal

surgery does not offer a significant increase of survival and should
be offered only when the patients are symptomatic (Table 2). However, it should be noted that mortality is very low in both modalities
of treatment.

In patients with double coronary artery disease the picture is
less clear. To give a more accurate impression we shall review
only recent results in which surgical technique has improved and in
which medical therapy can be considered completely up to date. It
can be seen in Table 3 that in most series the survival rate is
slightly better with bypass surgery. For better comparison, the
results of one older and two newer surgical series are given. Some
specific subsets have been recognized as being more dangerous, such
as combination of right coronary and left anterior descending artery
occlusion (Plattia et al., 1980).

Table 3. Medical and Surgical Treatment of Double
          Coronary Artery Disease

Authors	Number of Patients		Follow-up Survival % 5 years	
	M	S	M	S
Bruschke et al. 1973	233		62	
Kouchoukos et al. 1977	59	22	85	92
Hammermeister et al 1977	118	459	83	95
European Study Group 1980	154	147	87	92
Lawrie et al. 1978		390		90
Texas Heart Institute 1981		1399		90

Table 4. Medical and Surgical Treatment of Triple
          Coronary Artery Disease

Authors	Number of Patients		Follow-up Survival % 5 years	
	M	S	M	S
Bruschke et al. 1973	118		46	
Kouchoukos et al. 1977	66	250	58	85
Hammermeister et al. 1977	61	405	83.4	88.2
European Study Group 1980	188	219	84.7	95
Lawrie et al. 1978		342		86
Texas Heart Institute 1981		1723		89.6

Finally, in three vessel disease (Table 4) most authors agree
nowadays that survival can be improved with revascularization
(Bloomer and Ellestad, 1979). It is interesting that in Hammermeister's study (1977) no statistical difference was found between
groups, but as pointed out by the authors, their numbers were too
small to show appreciable differences (Hammermeister, 1981).

It should be noted that in newer series, more than three grafts
are consistently being reported, since atherosclerotic disease quite
often involves more than three coronary arteries. Multiple grafts
with complete revascularization does not add significantly to surgical mortality, thanks to current techniques. However, it does
significantly improve later survival. This was already evident in
1976 (Stiles et al.) in patients with triple coronary artery disease
and very recently reconfirmed by Cukignan et al. (1980). The experience at the Texas Heart Institute points this out quite clearly:
patients with multiple bypass grafts had improved survival (Table 5)
and more complete relief of symptoms (Figure 1).

Table 5. Texas Heart Institute - Coronary Bypass Surgery
          - Mortality Curves. October 1969 - December 1980

# Grafts	# Patients	Mortality		Actuarial Survival % Years			
		Hosp	Late	2	4	6	8
1	1412	2.4	5.0	96	92	90	82.5
2	4635	3.2	6.1	93	90	85.5	76
3	7602	3.6	3.9	93.6	90	86	79.5
4	4129	2.5	2.0	95	92	88	
5	879	1.9	1.4	96	93		
>6	82	3.7	2.4	92	92		

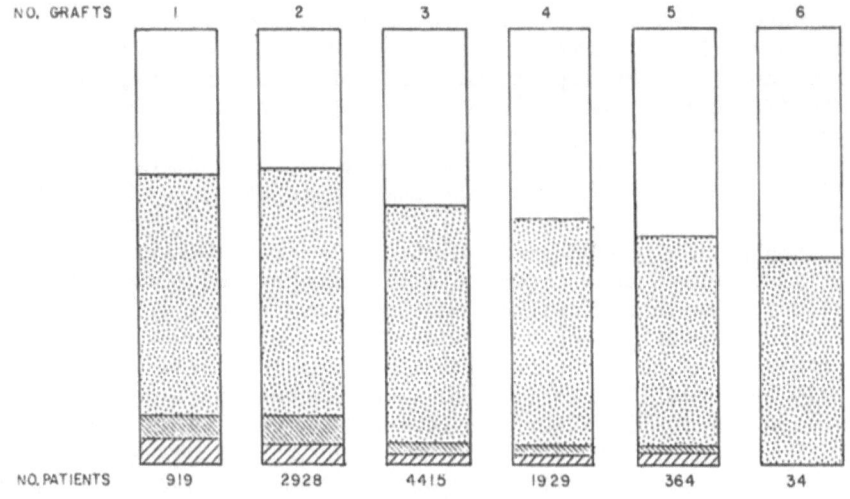

Fig. 1.   Texas Heart Institute – Isolated coronary bypass surgery:
Status of long-term survivors. 10/69 – 12/80

Here it should be noted that while authors may argue whether
surgery offers better survival, there is unanimous agreement that
it offers far greater relief of stable, chronic angina than medical
treatment (Hurst et al., 1978). Thus, Mathur and Guinn (1979) found
that at 4-7 years, only 5% of their patients with medical treatment
were asymptomatic, as compared to 50% of those with surgical treatment. Similarly, in the European Study Group (1980) only 20% of
those with medical treatment were free of angina as compared to 50%
with surgical.

Agreement is far from unanimous in several subsets of coronary
artery disease:

a) In patients with severe impairment of left ventricular
function, bypass surgery is accompanied by a higher mortality, and
no significant improvement in cardiac function is expected postoperatively. However, two aspects should be taken into account:
first, operative mortality has markedly declined during the last
five years, so older statistics do not necessarily apply today; and
second, with medical management the long-term survival of these
patients is dismal. Thus, if significant coronary artery disease

exists, surgery may be indicated, with the aim of preventing a future infarction which will irreversibly tip the scales against the patient. Actually, most centers report improved survival with surgery in these patients (Table 6). The decision for surgery is probably influenced by significant angina in this group. It should be mentioned that some authors believe that coronary artery bypass reduces the risk of sudden death, but others do not share this opinion (Table 7).

  b) In patients undergoing aortic valve replacement, it is currently suggested that coronary arteriography should be routinely carried out, and that if significant coronary artery disease is found, bypass surgery should be performed simultaneously with aortic valve replacement to decrease both early and long-term mortality. It should be noted that current surgical mortality is not appreciably higher with the combined surgery than with aortic valve replacement alone. However, as pointed out very recently by Bonow et al. (1981), it has not been proven that simultaneous bypass surgery will improve either short or long-term survival (Table 8). However, it is fair to admit that currently the trend is towards combined surgery.

  c) A similar situation exists when left ventricular aneurysm is present. Almost all surgeons currently perform bypass surgery simultaneously with aneurysm resection. A longer survival than with aneurysm excision alone is expected but not proven, although patients tend to have greater symptomatic improvement (Hall et al., 1976). At the Texas Heart Institute, experience with 1572 patients shows that combined operation is not associated with increased surgical mortality (Table 9).

Table 6. Results of Aortocoronary Bypass Surgery in Patients with Severe Left Ventricular Disfunction

Authors	No of Pts		Early Mortality %		Late Survival %		Follow-up Years
	M	S	M	S	M	S	
Yatteau et al. 1974	42	24		33	62	50	2
Isom et al. 1975		62		9.7		85	2
Manley et al. 1976	155	246		6.9	30	53	6
Vliester et al. 1977	30	56			68	90	3
Zubiate et al. 1977	79	140		9	20	58	5
Hung et al.		25		12		73	3

Table 7.   Influence of Coronary Artery Bypass
Surgery on Sudden Death

Authors	No of Pts		V.F. or Sudden Death %			Follow-up Months
	M	S	M		S	
Tilkian et al. 1976	47	44	14.9	11.3	NS	12
Vismaru et al. 1977	96	121	23.9	5.8p<.001		39
Hammermeister et al. 1977	350	1180	1.8-10.9 times higher			48
Mathur and Guinn 1979	60	56	5	9	NS	66

In this context we would like to present the experience at the
Texas Heart Institute of combining bypass surgery with other procedures (Table 10).

d)   Extensive investigation is applied to patients who have
already suffered an acute myocardial infarction. The current policy
is to subject them to a submaximal exercise test before they leave
the hospital, to detect significant arrhythmias or manifestations
of ischemia (Rackley et al., 1981). In the case of the former,
antiarrhythmic therapy is prescribed and its efficacy is systematically and serially assessed by Holter monitoring and subsequent
exercise tests. If ischemia becomes apparent, either before discharge or in a maximal test at six to twelve weeks after the acute
infarction coronary arteriography is recommended. Radionuclide
studies are currently considered valuable adjuncts.

e)   The outlook is gradually changing in impending myocardial
infarction (Table 11). Although intensive medical therapy is still
advocated as the initial treatment of choice, urgent surgical

Table 9.   Mortality of Patients with Aneurysmectomy only,
and with Combined Bypass Surgery

	No of Pts	Early Deaths %	Late Deaths %	
Aneurysmectomy only	221	7.69	4.52	NS
Combined with Bypass	1251	8.23	8.47	
Combined with Bypass and Septoplasty	421	8.79	6.89	NS

Table 8. Results of Combined Aortic Valve Replacement and Coronary Bypass Surgery

Authors	NORM AVR	AVR	CAD AVR+CAB	Mortality % Hospital NO CAD AVR	AVR	CAD AVR+CAB	TIME	Long term NO CAD AVR	AVR	CAD AVR+CAB
Brendt et al., 1974	40		28	0		14.3	24	10		0
Loop et al., 1977			80		9		42			35
MacManus et al., 1978	196		129	11		6.2	26	19	1	13.5
Richardson et al., 1979			220		5.4		36			23
Miller et al., 1979	139	31	101	2.2	8.0	8.3	36	5	10.3	6.5
Bonow et al., 1981	142	55		5	4		48	23	26	
Kirklin and Kouchoukos 1981	489		251	2	3.6					

Table 10. Texas Heart Institute Experience of Bypass Surgery
           with other Operative Procedures - 1969 - 1980

| Additional | No of Pts | | Mortality % | | | |
| | | | Early | | Late | |
Procedure	M	F	M	F	M	F
Aneurysmectomy	1352	181	7.5	11.6	8.0	10.5
Aortic Valve Surgery	639	100	12.8	19.0	8.5	9.0
Mitral Valve Surgery	396	170	13.6	13.5	10.9	11.8
Aortic + Mitral Valve Surgery	47	15	27.4	6.7	8.5	40.0

intervention is necessary in some patients (Brooks et al., 1981).
This reduction in risk, which is reflected in all aspects of coronary bypass surgery, has been achieved with better pre-operative
preparation and intra-operative myocardial preservation with cardioplegia and hypothermia and post-operative intra-aortic compensation.
Despite these advances, however, the place of surgery in the treatment of acute myocardial infarction remains unresolved. It should
be noted that Berg et al. (1981) operated upon 227 consecutive
patients with evolving acute myocardial infarction, at an average
of less than six hours after the onset of chest pain. Hospital
mortality was 1.76% and first year mortality 1.44%. while conventional treatment in 200 patients was associated with an in-hospital
mortality of 11.5%. Here it should be stressed that percutaneous
transluminal angioplasty, intracoronary thrombolysis with mechanical
and thrombolytic agents (Lee and Mason, 1981) may in the future
play a more important role than today.

     Thus one can safely say that the survival of the patient with
coronary artery disease has been improved, both by the introduction
of newer pharmaceutic agents, and by the judicious use of advanced
surgical techniques, and that coronary artery disease is associated
with a much more hopeful outlook than fifteen years ago.

Table 11.   Impending Myocardial Infarction – Unstable Angina
Effect of Bypass Surgery

| | Medical Management | | Surgical Management | | % Medical-Surgical |
	Acute	Long-term	Acute	Long-term	Management
Bertolasi et al. 1974	15	5.9	7.0	0	
Selden et al. 1975	0	5.5	4.8	0	4
Hultgren et al. 1977	4.5	16.6	1.9	3.9	12
National Co-operative Study–ST elevation during pain – 1978	2.7	6.8	4.9	4.9	36
Brooks et al. 1981	0	5.9	5.9	0	38.9

REFERENCES

Berg Jr., R., Selinger, S. L., Leonard, J. J., Grunwald, R. P.,
O'Grady, W. P., 1981, Immediate coronary artery bypass for
acute evolving myocardial infarction, J. Thorac. Cardiovasc.
Surg., 81:493-497.

Bloomer, W. E., Ellestad, M., 1979, Update on surgery for coronary
artery occlusive disease, Cardiovasc. Dis., 6:219-242.

Bonow, R. O., Kent, K. M., Rosing, D. R., Lipson, L. C., Borer, J.
S., McIntosh, C. L., Morrow, A. G., Epstein, S. E., 1981,
Aortic valve replacement without myocardial revascularization
in patients with combined aortic valvular and coronary artery
disease, Circulation, 63:243-251.

Brooks, N., Warnes, C., Cattell, M., Balcon, R., Honey, M., Layton,
C., Sturridge, M., Wright, J., 1981, Cardiac pain at rest:
Management and follow-up of 100 consecutive cases, Br. Heart
J., 45:35-41.

Conley, M. J., Ely, R. L., Kisslo, J., Lee, K. L., McNeer, J. R.,
Rosati, P. A., 1978, The prognostic spectrum of left main
stenosis, Circulation, 57:947-952.

Cukingnan, R. A., Carey, J. S., Wittig, J. H., 1979, Influence of
complete coronary revascularization on relief of angina, J.
Thorac. Cardiovasc. Surg., 79:188-193.

DeMots, H., Boncheck, L., Roesch, J., Anderson, R., Starr, A.,
Rahimtoola, S., 1975, Left main coronary artery disease, Am.
J. Cardiol., 36:136-141.

European Coronary Surgery Study Group, 1980, Prospective randomized
study of coronary artery bypass surgery in stable angina
pectoria, Lancet, 2:491-495.

Hall, R. J., Garcia, E., Wukasch, D. C., Reul, G. J., Sandiford,
F. M., Norman, J. C., Hallman, G. L., Cooley, D. A., 1976,
Long-term results of coronary artery bypass, Cardiovasc. Dis.,
Bull. Texas Heart Inst., 3:22-31.

Hammermeister, K. E., Derouev, T. A., Murray, J. A., Dodge, H. T.,
1977, Effect of aortocoronary saphenous vein bypass grafting
on death and sudden death: Comparison of nonrandomized medically
and surgically treated cohorts with comparable coronary disease
and left ventricular function, Am. J. Cardiol., 39:925-934.

Hammermeister, K. E., 1981, In what type of coronary disease patient
does aortocoronary bypass surgery improve survival?, Cardiovasc.
Rev. Rep., 2:589-598.

Hurst, J. W., King, S. B., Logue, R. B., Hatcher, C. R., Jones,
E. J., Craver, J. M., Douglas, J. S., Franch, R. H., Dorney,
E. R., Cobbs, B. W., Robinson, P. H., Clements, S. D., Kaplan,
J. A., Bradford, J. M., 1978, Value of coronary bypass surgery,
Am. J. Cardiol., 43:308-329.

Lee, G., Mason, D. T., 1981, Editorial: Percutaneous transluminal
coronary recanalization: A new approach to acute myocardial
infarction therapy with the potential for widespread application, Am. Heart J., 101:121-124.

Miller, D. C., Stinson, E. B., Oyer, P. E., Rossiter, S. J., Reitz,
    B. A., Shumway, N. E., 1979, Surgical implications and results
    of combined aortic valve replacement and myocardial revascu-
    larization, Am. J. Cardiol., 43:494-501.
Platia, E. V., Grunwald, L., Mellits, E. D., Humphries, J. O. N.,
    Griffith, L. S. C., 1980, Clinical and arteriographic variables
    predictive of survival in coronary artery disease, Am. J.
    Cardiol., 46:543-552.
Rackley, C. E., Russell Jr., R. O., Mantle, J. A., Rogers, W. J.,
    Papapietro, S. E., 1981, Modern approach to myocardial in-
    farction: Determination of prognosis and therapy, Am. Heart J.,
    101:75-85.
Stiles, Q. R., Lindesmith, G. G., Tucker, F. L., Hughes, R. K.,
    Meyer, B. W., 1976, Long-term follow-up of patients with
    coronary artery bypass grafts, Circulation, 54:111-32.

DISSECTING AORTIC ANEURYSMS: PROSPECTS OF

MEDICAL AND SURGICAL THERAPY

Vincenzo Gallucci

Department of Cardiovascular Surgery
University of Padova Medical School
Padova, Italy

Acute dissection of the thoracic aorta is a lethal condition[1] demanding urgent and well-planned treatment, once the diagnosis is made with certainty. In fact about 90% of the patients die within one week from onset, usually because of hemorrhage, at times with cardiac tamponade, or with visceral ischemia and infarction or acute aortic insufficiency and heart failure. Ten per cent of the patients, spontaneously survive to the 'chronic' stage but most will die of aneurysmal rupture.

During the past 20 years, the better knowledge of its clinical manifestation and the easier access to sophisticated diagnostic facilities brought up to almost 100% the number of cases where a correct diagnosis was made in the hospitalized patient.[2] Therefore, the main problem remains the choice of the most appropriate therapy, be it medical or surgical, and its quick employment in each case.[3-6]

In a 12 year period (1970-1981), at the Department of Cardiovascular Surgery, University of Padova Medical School, 132 cases of dissecting aneurysm of the thoracic aorta were observed and treated. 33 of them were treated by medical means only, through the diagnostic phase and beyond, following Wheat's method[7-9] of reducing both mean systemic arterial pressure and dP/dT, with trimethaphan, reserpine and guanethidine, aiming at the conversion of the acute dissection into a subacute or chronic process. In this way better tissue would be available for the surgical intervention which was meant in each case, by reducing the edema and fragility of the aortic wall, developing fibrotic processes and stabilizing the lesion. Other pharmacologic agents were also employed during that phase, such as sodium nitroprusside[10], beta-blockers, diuretics, etc.

The medical (hypotensive) treatment was used in each patient under close medical and nursing surveillance, continuous monitoring of all vital parameters and readiness of operating room facilities. Unfortunately 32 out of 33 such treated patients died, mostly from rupture of the aneurysm or from complete heart block, or the consequences of cerebral and visceral aortic branch occlusion. Ninety-nine other patients instead underwent surgical treatment, once the diagnosis was established, mostly on an emergency basis. In my experience, medical treatment alone has never been curative and did not prevent the death of the patient. Furthermore, there are some well known instances where the hypotensive therapy is contraindicated These are shock, persistence of pain, meaning that the treatment does not keep the dissection from progressing, symptoms of cerebral or renal ischemia, signs of coronary insufficiency, ischemia of the limbs, acute aortic insufficiency.

It must be said, however, that there are also some unfavorable preoperative factors which worsen considerably the surgical prognosis. Among them, again, acute aortic insufficiency and heart failure, coronary occlusion and myocardial infarction, rupture of the aortic wall especially if associated with pericardial tamponade, neurologic damage, acute renal insufficiency. These last two complications, if persistent, contraindicate surgery as they are associated with 100% hospital mortality.

The operations performed, different according to the type of dissection, to the site of intima laceration, and to the peripheral extension of the hematoma, were the 'classical' ones proposed by De Bakey.[11] In every instance a portion of dissected aorta was resected including the 'entrance' to the false lumen and replaced with a Dacron prosthesis, obliterating the false lumen as carefully and completely as possible. Extracorporeal circulation, moderate hypothermia, cardioplegia were used with techniques and degrees varying with time and better knowledge of cardiac metabolism and physiology.

The two ways of treatment gave quite different results, statistically highly significant with a hospital mortality for the medical treatment alone, as said, close to 100%. The mortality after surgery, very high during the first years, has steadily declined, being about 25% in 1980-81. Medical (hypotensive) therapy as the only treatment of the acute dissecting aneurysm has therefore been abandoned at our Center.

Furthermore, the constant improvement of our results with the surgical therapy, due certainly to a better surgical technique but also to the widespread experiences with a more complete myocardial protection (hypothermia and cardioplegia) and safer arterial prostheses (low porosity Dacron), made us adopt the following current policy:

a) each patient suspected of having an acute dissecting aneurysm undergoes early intensive medical treatment (either hypotensive and/or supportive) during the diagnostic phase.

b) once the diagnosis is established, with special reference to the site of initial dissection (ascending or descending thoracic aorta)[12], the status of the coronary arteries[13], the visceral aortic branches and the limbs, the patient is taken to surgery without delay.

A good number of associated procedures were performed in the 99 operated patients, mostly in type I and II dissection (Tables I, II and III), the aortic valve being replaced numerous times. In fact, all kinds of conservative procedures on the acutely insufficient valve gave unsatisfactory results. Most recently, several patients underwent aortic valve replacement and reimplantation of the coronary arteries on the prostheses, with good outcome: these are not included in the present series. All of the operated patients who were discharged from the hospital have been followed regularly and many had repeated catheterization and angiography.

From the clinical point of view, some interesting observations have been made:

1) the actuarial survival at 12 years is about 91% which seems to be quite satisfactory, indeed, for a disease usually lethal if untreated.
2) the clinical conditions of the survivors are presently excellent and most of them are asymptomatic 1 to 10 years postoperatively.
3) as far as the arterial hypertension is concerned, all of the hypertensive patients remained such, despite adequate treatment.

Table I.   Dissecting Aneurysm Type I
Surgical treatments: 41 pts

Replacement of ascending aorta (RAA)	18	(15)
RAA + aortic valvuloplasty	5	(5)
RAA + aortic valve replacement (AVR)	6	(1)
RAA + aorto-coron. bypass (ACB)	1	(1)
RAA + AVR + ACB	1	
Bentall	4	(2)
Bentall + mitral valve repl.	1	
Intraluminal prosthesis	1	
Intraluminal prosth. + aortic valvulopl.	3	
Impossible aortic cannulation	1	(1)
Total	41	(24)

( ) Operative deaths

Table II.   Dissecting Aneurysm Type II
Surgical treatment: 25 pts

Replac. ascending aorta	6	(3)
RAA + AVR	10	(2)
RAA + AVR + MVR	1	
RAA + AVR + MVR + ACB	1	(1)
RAA + ACB	2	
Bentall	3	
RAA + AVR + ACB	1	
Patch closure of aortic tear	1	(1)
Total	25	(7)

( ) Operative deaths

4) three patients died during the 12 year follow-up period, one of
them after reoperation for aortic insufficiency and two others of
unknown causes.

However, these encouraging clinical findings are only partially
in agreement with those of cardiac catheterization and angiography
which show that only about 60% of the patients show a 'normal'
appearing thoracoabdominal aorta and branches. The remaining 40%
presented some very interesting features such as aneurysmal dilation of the sinuses of Valsalva, aortic insufficiency, further
dilation of the distal false lumen, new dissection, compression of
the true lumen and occlusion of previously open aortic branches.

There must be no doubt therefore that the acute episode of
dissection represents only one phase of the natural history of a
profoundly altered aortic wall and this is proved by the facts that:
a) the aortic root, left behind at the time of resection of the
ascending aorta in type I and II, has the tendency to dilate and
become aneurysmatic, thus involving also the coronary ostia and the
aortic valve.

Table III.   Dissecting Aneurysm Type III
Surgical treatment: 33 pts

Replac. descending aorta (RDA)	24	(7)
RDA + Ao-subclavian bypass	3	(1)
RDA + Ao-femoral bypass	1	
Intraluminal prosthesis	5	(2)
Total	33	(10)

( ) Operative deaths

b) the valve itself, even when initially intact and well functioning, can become insufficient not only because of annular dilatation but also with degeneration of the cusp tissue.
c) despite any attempt to obliterate it, the false lumen, beyond the distal anastomosis of the aortic prosthesis, can stay open or become further dilated, giving origin to a fusiform aneurysm;
d) a true new dissection can originate at the site of distal anastomosis with new intimal laceration.
e) the true aortic lumen can be compressed from outside by thrombi or hematoma present in the false lumen, so as to be stenosed.
f) both the thoracic and abdominal aortic branches can become occluded by thrombosis, intimal flaps, etc.

Little correlation can therefore exist sometimes between the clinical status of the operated patient which can be very satisfactory and his anatomic situation, potentially most dangerous.

As a consequence of the above observations, a few changes in our surgical technique have been made and routinely applied, hoping they might contribute to change the spontaneous evolution of these patients. As far as the distal suture line is concerned, where new dissection, intimal laceration or aneurysmal dilatation has been shown to occur, we now reinforce it from the outside with a tight band made of a flat teflon felt, pledgeting, etc.

In a good many cases of type I and II dissections, consideration is given to a routine replacement of the aortic valve and reimplantation of the coronary arteries ostia, by utilizing premade ascending aorta prostheses mounted with a mechanical valve. This is aimed at preventing both the dilation of the residual aortic root as well as aortic regurgitation, causing reoperation in a few years. In our recent series, this last procedure did not increase surgical mortality.

A recent addition to the surgical armamentary are the 'endoluminal' aneurysmal prostheses, made to eliminate the difficulties of suturing the fragile dissected aortic wall. They are inserted into the lumen and securely tied from outside. Although their use seems to be much more practical in the descending aorta than in the ascending, too proximal to the coronary ostia and easily compromising the motion of the aortic valve cusps, in some cases they proved to be useful and saved considerable cardiac arrest time.

REFERENCES

1. A. E. Hirst, Dissecting aneurysm of the aorta: A review of 505 cases, Medicine (Baltimore), 34:217 (1958).
2. P. G. Cévese, Aneurismi dissecanti dell'aorta, Piccin Ed., Padova (1977).

3. P. O. Daily, Management of acute aortic dissection, <u>Ann. Thorac.</u> <u>Surg.</u>, 10:237 (1970).
4. V. Gallucci, Aneurismi dissecanti dell'aorta: Indicazioni chirurgiche e terapia, VII Congr. Naz. Soc. It. Chir. Urg. e Pronto Socc.
5. J. N. Kidd, Surgical treatment of aneurysms of the ascending aorta, <u>Circulation</u>, 54:6 (1976).
6. W. G. Wolfe, The evolution of medical and surgical management of acute aortic dissection, <u>Circulation</u>, 56:4 (1977).
7. M. W. Wheat, Treatment of dissecting aneurysm of the aortia without surgery, <u>J. Thorac. Cardiovasc. Surg.</u>, 50:364 (1965).
8. M. W. Wheat, Surgical treatment of aneurysm of the aortic root, <u>Ann. Thorac. Surg.</u>, 12:593 (1971).
9. M. W. Wheat, Dissecting aneurysm of the aorta: Gibbon surgery of the chest, Saunders Ed., Philadelphia (1977).
10. P. Siegel, Sodium nitroprusside in the surgical treatment of cerebral aneurysm and arteriovenous malformation, <u>Br. J.</u> <u>Anaest.</u>, 43:790 (1971).
11. M. E. De Bakey, Surgical management of dissection, <u>Ann. Thorac.</u> <u>Surg.</u>, 10:237 (1965).
12. A. Appelbaum, Ascending vs descending aortic dissection, <u>Ann.</u> <u>Surg.</u>, 21:298 (1976).
13. P. Zubiate, Surgical treatment of aneurysms of the ascending aorta with aortic insufficiency and marked displacement of the coronary ostia, <u>J. Thorac. Cardiovasc. Surg.</u>, 71:415 (1976).

THE ROLE OF CARDIAC PACING IN HEART BLOCK

COMPLICATING ACUTE MYOCARDIAL INFARCTION

A. Raineri, G. Mercurio, B. Candela,
G. L. Piraino, and P. Assennato

Cattedra di Fisiopatologia Cardiovascolare
Università di Palermo
Italy

The organization of the Coronary Care Unit and the use of electrocardiographic monitoring have allowed a re-evaluation of the incidence of the atrioventricular conduction disturbances in acute myocardial infarction.

In the opinion of various authors[1-8] such incidence ranges between 3 and 21%. Nevertheless it is appropriate to consider that many of these studies were carried out in times when the methods of observation did not permit a continual revelation of the electrical events of the heart. Recent studies in fact show on average a higher incidence of atrioventricular conduction disturbances. This has allowed the intervention at the earliest time in cases that need cardiac pacing.

Nevertheless, it is necessary to add that cardiac pacing is not considered an intervention to utilize in every circumstance, but it is proposed only in cases in which there is a danger of sudden heart block or severe bradycardia. In fact it seems that the risk of endocardial pacing in the course of an acute myocardial infarction must not be neglected[9].

The appearance of the atrioventricular block during the acute stage of an infarction indicates that some structures of the conduction have been affected in an elective way, but it can also indicate that it happened as a result of an extension of the necrotic process. These different situations show different pathological patterns that have different prognostic significance[30].

The purpose of this study is to evaluate the role that cardiac
pacing may carry out in the course of acute myocardial infarction
complicated by atrioventricular block and more precisely if the introduction of this technique has influenced the patient's prognosis
in a short and long period.

MATERIAL AND METHODS

Our cases include 442 patients with acute myocardial infarction
admitted to the C.C.U. from January 1976 to February 1982. Of these
353 were men and 89 were women with an average age of 59 and 63
respectively.

Acute myocardial infarction was diagnosed on the basis of clinical symptoms, variation of enzymes index of myocardial necrosis,
typical electrocardiographic pattern.

In the presence of 2° and 3° AV block and 1° AV block associated
with intraventricular conduction disturbances a temporary pacemaker
was inserted. A permanent pacemaker was inserted when the above
mentioned conditions continued after the acute phase of infarction.

For statistical analysis the chi-square Pearson test was utilized to compare frequences, and student's "t" test to compare the
means.

RESULTS

Of 442 patients with acute myocardial infarction, 238 (53.8%)
had a necrosis localized in the anterior wall and 204 (46.2%) in
inferior wall (Table 1).

In 69 patients (15.6%) the course of the illness was complicated
by the disturbances of atrioventricular conduction. 17 were women
with an average age of 53 (52-81 years); and 52 men, with an average
age of 64 (45-84 years). The AV block was more frequent in the inferior infarction (74%) than the anterior infarction (26%).

In 55 patients a temporary pacemaker was inserted. This means
that 79.7% of patients with AV block need cardiac pacing. This was
necessary in 12 patients with anterior infarction (21.8%) and in
43 (78.2%) with inferior infarction. In 14 patients this measure
was not used because of the fugacity of the complication.

In 16 patients (6 with anterior myocardial infarction and 10
with inferior myocardial infarction), because of the persistence
of the atrioventricular block after the acute phase a permanent pacemaker was inserted.

Table 1.   Confirmed Acute Myocardial Infarction
(A.M.I.) Admitted in C.C.U. from January
1, 1976 to February 28, 1982: Site of
M.I. Incidence of Atrio-ventricular
(A-V) Heart Block and Number of Tempor-
ary Cardiac Pacing (T.C.P.) and
Permanent Cardiac Pacing (P.C.P.)

	Anterior M.I.		Inferior M.I.		Total	
	n.	%	n.	%	n.	%
A.M.I. admitted in C.C.U.	238	53.8	204	46.2	442	100
A.M.I. without A-V heart block	220	59.0	153	41.0	373	84.4
A.M.I. with A-V heart block	18	26.0	51	74.0	69	15.6
A.M.I. with A-V block and T.C.P.	12	21.8	43	78.2	55	79.7
A.M.I. with A-V block and P.C.P.	6	37.5	10	62.5	16	23.2
A.M.I. with A-V block no paced	6	42.8	8	57.2	14	20.3

A complete atrioventricular block was more frequent (60%) than
the 1° AV block (10.9%) and the 2° (29.1%). AV Block is more frequent (78.7%) in inferior myocardial infarction than in the anterior
one (21.8%). (Table 2).

Table 3 shows hospital mortality. No significant statistic
difference exists among the three following groups: group with infarction without AV block, group with infarction and AV block and
the final group with infarction and AV block in which a pacemaker
was inserted. There was no difference in mortality in any of the
groups also if the infarction was in anterior or inferior wall.

In the group with a very transient atrioventricular block, in
which pacing was not necessary, no mortality was observed.

Table 2.   Type of A-V Heart Block Complicating A.M.I.
in 55 Paced Patients

A-V HEART BLOCK	Anterior M.I.		Inferior M.I.		Total	
	n.	%	n.	%	n.	%
I A-V heart block with intraventricular - blocks	2	33.3	4	66.9	6	10.9
II A-V heart block	2	12.5	14	87.5	16	29.1
Complete A-V heart block	8	24.2	25	75.8	33	60.0
Total	12	21.8	43	78.2	55	100

Table 3. Hospital Mortality Rate in A-V Block
Complicating A.M.I. with and without
Temporary Cardiac Pacing (T.C.P.)

		Total No. of A.M.I		Hospital mortality	
		n.	x	n.	x
A.M.I. without A-V block	Anterior	220	—	32	14.5
	Inferior	153	—	23	15.0
A.M.I. with A-V block	Ant.	18	—	3	16.6
	Inf.	51	—	7	13.7
A.M.I. with A-V block and T.C.P.	Ant.	12	—	3	25.0
	Inf.	43	—	7	16.2
A.M.I. with A-V block without T.C.P.	Ant.	6	—	—	—
	Inf.	8	—	—	—

Table 4 shows the late mortality. The group with permanent
pacemakers shows a higher mortality rate (37.5%) than the group
with temporary cardiac pacing (T.C.P.) (10.7%) and without T.C.P.
(7.1%). The differences are significant (p 0.05). In this same
group the mortality was higher in the anterior infarction (66.6%)
than the inferior (20.0%). The difference is not significant.

Table 5 shows the cause of death and average mortality during
hospitalization in the group of 69 patients with acute myocardial
infarction and AV block. The cause of death in patients with
necrosis localized in the anterior wall was cardiogenic shock (3
patients) with average survival of 39 hours. In the patients with
inferior infarction there were 5 deaths caused by cardiogenic shock
and 2 by left ventricular rupture, with average survival of 11 days.

Table 6 shows the cause of late mortality in 58 patients with
acute myocardial infarction and AV block. 16 patients in this group
had a permanent pacemaker.

Table 4. Late Mortality Rate in A-V Block
Complicating A.M.I. with and without
Temporary Cardiac Pacing (T.C.P.) and
with Permanent Cardiac Pacing (P.C.P.)

		Total No. of A.M.I		Late mortality		
		n.	x	n.	x	
A.M.I. with A-V block	Anterior	15	—	6	40.0	17.2
	Inferior	43	—	4	9.30	
A.M.I. with A-V block and T.C.P.	Ant.	3	—	1	33.3	10.7
	Inf.	25	—	2	8.00	
A.M.I. with A-V block without T.C.P.	Ant.	6	—	1	16.6	7.1
	Inf.	8	—	—	—	
A.M.I. with A-V block and P.C.P.	Ant.	6	—	4	66.6	37.5
	Inf.	10	—	2	20.0	

Table 5. Average Hospital
Mortality Time and
Cause of Death among
69 Patients with A-V
Heart Block Compli-
cating A.M.I.

CAUSE OF DEATH	Anterior M.I.	Inferior M.I.
Shock	3	5
L.V. Rupture	—	2
Time of death	39 hours	11 days

Of the total of 6 deaths in patients with anterior wall infarction, 4 deaths were caused by cardiogenic shock (of which 2 with
permanent pacemakers inserted) and 2 by sudden death (both of them
had permanent pacemakers). The average survival in the 2 groups
was 32 months in permanent pacemaker patients, and 7.5 months in
the other group.

Finally in the inferior infarction the 4 deaths were caused
by a reinfarction: 2 patients had permanent pacemakers. The average
survival was 19 months in patients with pacemakers and 20 months
in patients without pacemakers.

In the group between 60-69 we find more myocardial infarction,
but in the age group 70-79 there are more AV block (26.3%).(Figure 1).

Table 7 compares the localization of the infarction, the duration of QRS and the mortality rate, in 55 patients that had had a
temporary pacemaker. In anterior infarction 9 patients showed a

Table 6. Late Mortality in 58 A-V Block
Complicating M.I. Patients with and
without Permanent Cardiac Pacing (P.C.P.)

CAUSE OF DEATH	M.I. with A-V block and P.C.P. n.=16		M.I. with A-V block without P.C.P. n.=42	
	Anterior M.I.	Inferior M.I.	Anterior M.I.	Inferior M.I.
Shock	2	—	2	—
Sudden death	2	—	—	—
Re-infarction	—	2	—	2
Time of death	32 months	19 months	7.5 months	20 months

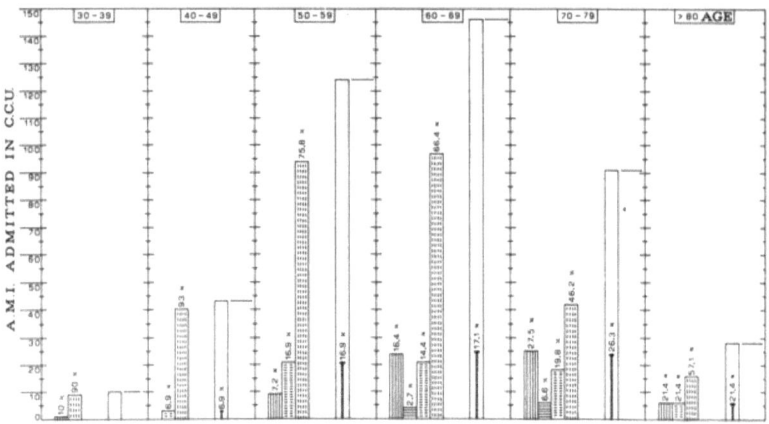

Fig. 1. Age distribution of atrio-ventricular (A-V) heart block,
         complicating A.M.I. and mortality rate in 442 patients
         admitted in C.C.U. ▦ Dead without A-V block; ▦ dead
         with A-V block; ☐ number of patients; ▦ alive without
         A-V block; ▦ alive with A-V block; ┃ A-V heart block
         incidence.

narrow QRS and in this group there have been 2 deaths. 3 patients
showed a wide QRS and one of them died. In inferior infarction 35
patients showed a narrow QRS and 5 died. 8 patients showed a wide
QRS and 2 died. Statistical analysis has not shown any significant
difference.

    Last of all the rate of enzyme index of the myocardial necrosis
was considered. The various groups examined with or without pace-

Table 7. Location of Infarction and QRS
          Duration Related to Mortality in 55
          Patients with Temporary Pacemakers

	Anterior M.I.		Inferior M.I.	
	n.	%	n.	%
Total number of patients	12	—	43	—
NARROW QRS	9	75.0	35	81.4
Mortality	2	22.2	5	14.3

	Anterior M.I.		Inferior M.I.	
Total number of patients	12	—	43	—
WIDE QRS	3	25.0	8	18.6
Mortality	1	33.3	2	25.0

makers, with anterior or inferior infarct, alive or dead did not
have a significant difference.

DISCUSSION

In our material of 442 patients with acute myocardial infarction
we had 69 cases (15.6%) complicated by atrioventricular block.

51 of these (74%) were infarction of the inferior wall. These
results agree with other authors[6,8-10,14,16,17].

The highest incidence of atrioventricular block in inferior
infarction depends on the occlusion of the vessels that supply blood
to the AV node. In fact in 80% of the cases the AV node is perfused
by the descending branch of the right coronary[10-12]: the occlusion
of this vessel is responsible in 90% of the cases for the inferior
wall infarction. The low percentage (26%) of anterior wall infarction complicated by atrioventricular block, may be justified by the
fact that the anterior descending artery, is not responsible for
the blood supply to the atrioventricular node.

The incidence of atrioventricular block in the anterior myocardial infarction may be an expression of a large necrosis[4,13,14]
or the contemporary involvement of more vessels or both.

In our cases we have not found any difference in mortality
between acute myocardial infarction complicated by atrioventricular
block and those not complicated by atrioventricular block.

Therefore we can conclude that the appearance of the atrioventricular block during acute myocardial infarction does not constitute
an unfavourable prognosis in comparison to the infarction without
atrioventricular block.

A different opinion was expressed by some authors[4,6,13,14,18]
who have reported in their cases a higher mortality in acute myocardial infarction complicated by atrioventricular block, and in
particular if it was an anterior wall infarction.

Stock and Norris report a mortality of 80% in anterior infarction: Gupta, 28% in inferior infarction. There are some who assert
(Ginks 1977) that a lower percentage of mortality may be the result
of some selection of cases by their ability of surviving transfer
to hospital[13].

It is in fact noted that if the atrioventricular block compli-
cates precociously the acute phases of infarction, it is an extremely
unfavourable condition, such as to increase mortality[6].

We must state that our patients do not come from other hospitals and that they have been admitted to C.C.U. within the first 12 hours from the beginning of pain.

So it is possible to point out two hypothesis:

(1) that cardiac pacing has a preventive action;
(2) that the use of drugs, that are now more widely used with the purpose of limiting the area of infarction also have a favourable effect on these patients.

The indication of temporary cardiac pacing could come from the knowledge of the site of the conduction block rather than to its degree[24]. The distinction is possible on the basis of electrophysiologic studies. Since the electrophysiological study is extremely dangerous in the acute phase of an infarction, it is more advisable to follow the clinical condition of patients in order to decide on cardiac pacing.

We are particularly prone to cardiac pacing having utilized this procedure in 79% of the cases.

As we have already pointed out, this might be one of the reasons why the mortality verified in patients with acute myocardial infarction complicated by atrioventricular block is not different from those patients with acute myocardial infarction not complicated by atrioventricular block.

Seeing the opinions of the various authors the results somewhat disagree: there are those who insert a temporary pacemaker in all cases of AV conduction disturbances[6,8,13,23]; while there are others who do not insert it preventively[14,15,26,27].

The percentage of mortality in both groups is similar. Rather, the mortality rate reported by Jackson who has not utilized preventive cardiac pacing, is among the lowest[20].

Here, it seems important to refer to the experience of a group that treated acute infarction with or without the insertion of pacemakers. In the patients with pacemakers the mortality was remarkably different in relation to:

(1) use of fixed rate pacemakers (56%);
(2) use of demand pacemakers in the first years of their experience (38%)
(3) use of demand pacemakers in the last years of their experience (11%).

They themselves refer that the factors may be responsible for the probable improvement were thought to be:

(1) early pacing with avoidance of syncope;
(2) improvement in cardiac output;
(3) demand pacing with awareness of possibility of inappropriate
    stimulation;
(4) free use of suppressant drugs.

Therefore, in accordance with literature[8] we must conclude that
the insertion of a temporary pacemaker in the phase of acute myocardial infarction complicated by atrioventricular block is very
useful[8,9].

Moreover, it is stressed that anterior infarction has a more
unfavourable prognosis in respect to inferior acute myocardial infarction when it is complicated by AV block. It is said in fact
that the anterior myocardial infarction rarely gives conduction
disturbance, which, if present, has nevertheless a more severe prognosis[10]. But our experience is different. In fact we did not find
any mortality difference between these two groups of patients.

It is important to note that the involvement of the conduction
tissues is not only to be attributed to necrosis[30], but to:

(a) transient ischemia;
(b) inappropriate vagal tone;
(c) edema and inflammation;
(d) vascular spasm.

When these conditions regress the AV conduction disturbances
disappear. For this reason most of the patients who underwent cardiac pacing did not need to have permanent pacemakers inserted.
This in fact was verified only in 16 patients out of 69.

We have also carried out a follow-up on temporary paced patients
and we have been able to note that these patients did not need permanent pacing. This shows that only necrosis of the conduction system
makes permanent pacing necessary; which, however, does not improve
the prognosis. In fact if we observe permanent paced patients after
hospital discharge, we note that a high rate of mortality exists
among them.

To conclude:

(1) There is no significant mortality rate difference between in-
    ferior and anterior acute myocardial infarction complicated by
    atrioventricular block, in our experience.
(2) Temporary pacing prevents the risk associated with asystolia,
    low cardiac output and allows the use of drugs, which otherwise
    should be avoided.
(3) Besides necrosis, some reversible pathological conditions might
    give conduction disturbances.

(4) The atrioventricular block is considered an unfavourable complication only when it is determined by necrosis: this is a condition that makes the block irreversible.

SUMMARY

In 442 cases of acute myocardial infarction, 69 were complicated by atrioventricular block. Of these 51 (74%) were inferior infarction.

We know that most of the incidence of the atrioventricular block in the inferior infarction depends on the occlusion of vessels that supply blood to the AV node. Moreover, the incidence of atrioventricular block in the anterior infarction might depend on the large necrosis or the involvement of more vessels or both.

In our experience no difference of mortality was shown between anterior and inferior acute myocardial infarction complicated by atrioventricular block.

The low death rate verified in our cases makes us conclude that the insertion of a temporary pacemaker in the acute myocardial infarction complicated by atrioventricular block is extremely useful. When the atrioventricular block does not regress, ie. a sign of conduction tissue destruction.

Such an event is to be considered prognostically unfavourable. In fact in the group of patients in which it was necessary to change from temporary to permanent pacemakers because of persistence of the atrioventricular block the death rate was high.

REFERENCES

1. V. Alexandrow and P. Borkowsk, Dyderszynsky:Blok przedsionkowo komorowj wprzebiegn zewaln serca: deswiadezema wlasen, Kard. Pol. 21:253 (1978).
2. R. W. Brown, D. Hant and J. G. Sloman, The natural history of atrio ventricular conduction defects in acute myocardial infarction, Am. Heart J. 44:460 (1969).
3. P. T. Errazquin, March I.:Atrioventricular block in acute myocardial infarction, Acta Cardiol. 25:217 (1970).
4. M. C. Gupta, M. M. Singh, P. K. Wahal, M. P. Mehrotra and S. K. Gupta, Complete heart block complicating acute myocardial infarction, Angiology 29:749 (1979).
5. D. B. Hackel and E. M. Estes, Pathologic features of atrioventricular and intraventicular conduction disturbances in acute myocardial infarction, Circulation 43:977 (1971).

6. W. J. Kostuk and D. S. Beanlands, Complete heart block associated
    with acute myocardial infarction, Am. J. Cardiol. 26:380
    (1970).
7. R. M. Narvas, J. M. Kilgour and S. K. Basu, Heart block in acute
    myocardial infarction: prognostic factors and role of trans-
    venous catheter pacemaker, Canad. Med. Ass. J. 102:55 (1970).
8. C. Di Jorio, P. R. Marulli, C. Corona, I. Leonzio and A. Rossi,
    La stimolazione elettrica endocavitaria nei blocchi A-V com-
    plicanti la fase acuta dell'infarto miocardico, Min. Med. 71:
    2151 (1980).
9. K. Chatterjee, A. Harris and A. Leatham, The risk of pacing
    after infarction and current recommendation, Lancet 15:1061
    (1969).
10. G. Borello, E. Bellone, G. C. Bulgarelli, A. De Bernardi,
    R. Bevilaqua, A. Pizzuti, A. Alvino, M. Abrate and
    P. F. Angelino, Indicazione all'elettrostimolazione temporanea
    in corso di infarto miocardico acuto, Min. Cardioangiol. 29:
    137 (1981).
11. T. N. James and G. E. Burch, Blood supply of the human intra-
    ventricular septum, Circulation 17:391 (1958).
12. T. N. James and G. E. Burch, The atrial coronary arteries in
    man, Circulation 17:90 (1958).
13. W. R. Ginks, R. Sutton, OHW. and A. Leatham, Long-term prognosis
    after acute anterior infarction with atrioventricular block,
    Br. Heart J. 39:186 (1977).
14. D. B. Cohen, L. Doctor and A. Pick, The significance of atrio-
    ventricular block complicating acute myocardial infarction,
    Am. Heart J. 55:215 (1958).
15. S. R. Courter, J. Moffat and N. O. Fowler, Advanced atrioventri-
    cular block in acute myocardial infarction, Circulation 27:
    1034 (1963).
16. E. A. Paulk and J. W. Hurst, Complete heart block in acute myo-
    cardial infarction, Am. J. Cardiol. 17:695 (1966).
17. J. Beregovich, S. Fenig, J. Lasser and D. Allen, Management of
    acute myocardial infarction complicated by advanced atrio-
    ventricular block, Am. J. Cardiol. 25:54 (1969).
18. R. M. Morris, Heart block in posterior and anterior myocardial
    infarction, Brit. Heart J. 31:352 (1969).
19. M. E. Scott, J. S. Geddes and G. C. Patterson et al., Management
    of complete heart block complicating acute myocardial infarc-
    tion, Lancet 2:1382 (1967).
20. M. De'Thomatis, R. Gallesio, G. Musso and N. Falciola, La stimo-
    lazione ellettrica ventricolare temporanea del cuore, Boll.
    Soc. It. Cardiol. 23:847 (1978).
21. T. N. James, Anatomy of the coronary arteries in health and
    disease, Circulation 32:1020 (1965).
22. P. F. Fazzini, P. Larchi and P. Pucci, Il significato prognostico
    dei blocchi intraventricolari nell'infarto miocardico acuto,
    Giorn. It. Cardiol. 5:526 (1975).

23. M. Rotman, S. G. Wagner and A. G. Wallace, Bradyarrhythmias in
    acute myocardial infarction, Circulation 45:703 (1972).
24. O. S. Narula, B. J. Sherlag and R. P. Janer, Analysis of AV con-
    duction defect in complete heart block utilizing His bundle
    electrograms, Circulation 41:437 (1970).
25. O. S. Narula and P. Samet, Wenckebach and Mobibitz II Block due
    to lesion within the His bundle and bundle branches,
    Circulation 41:947 (1970).
26. C. K. Fridberg, H. Cohen and E. Donoso, Advanced heart block
    as complication of acute myocardial infarction: Role of
    pacemaker therapy, Prog. Cardiovasc. Dis. 10:466 (1968).
27. W. E. Jackson and F. A. Bashour, Cardiac arrhythmias in acute
    myocardial infarction: 1. Complete heart block and its
    natural history, Dis. Chest 51:31 (1967).
28. M. Prinzmetal, R. Kenuamer, R. Merliss et al., Angina pectoris:
    1. A variant form of angina pectoris, Am. Med.27:375 (1959).
29. R. Botti, A variant form of angina pectoris with recurrent trans-
    ient complete heart block, Am. J. Cardiol. 17:443 (1966).
30. P. K. Gupta, E. Lichstein and K. D. Chadda, Heart block compli-
    cating acute inferior wall myocardial infarction, Chest 69:599
    (1976).
31. S. A. Forsberg and S. Juul-Möller, Myocardial infarction compli-
    cated with heart block: treatment and two-years prognosis,
    Pace 2:49 (1979).

HOW TO PREVENT HEART FAILURE ?

W. Rutishauser

Center of Cardiology
University Hospital
Geneva, Switzerland

PATHOPHYSIOLOGY OF AND COMPENSATORY MECHANISMS IN HEART FAILURE

Heart failure is a complex pathophysiologic disorder, which can be defined as the inability of the heart to pump blood at a rate commensurate with the ordinary metabolic demands of the organs and tissues in spite of sufficient venous return. Although the basic alterations lies in the myocardium, the obvious clinical abnormalities present in congestive heart failure are mostly due to regulatory changes and overshooting compensatory mechanisms[1].

The simple clinical signs of overt congestive heart failure are due to an increased filling pressure in an enlarged heart with decreased ejection fraction and limited output which demonstrate the reduced contractile reserve.

While the diagnosis of overt congestive failure is simple, recognition of early heart failure needs skill or precise measurements. In doubt, we need a stress test to have a firm diagnosis. While a normal heart increases its output under exercise stress with an unchanged or even lower filling pressure, an abnormal heart shows an increase in end-diastolic pressure when stroke volume is increased and in severe heart failure with strongly depressed function, end-diastolic pressure rises markedly while stroke volume falls eventually considerably.

If we compare the pressure-volume loops at rest and during exercise of a normal untrained heart, a well-trained heart of a marathon runner and a heart in chronic failure, the end-diastolic

volume is at rest may be similar in the latter two, however the stroke
volume and the contractile reserve of the trained person exceeds by
far the normal and, naturally, that of the failing heart which cannot
increase its ejection fraction under stress.

The interrelationship of different parameters in heart failure
is shown in Figure 1. The reduced contractility, together with a
severely increased afterload in congestive heart failure, decreases
markedly the extent of fiber shortening, so that the stroke volume
is reduced. Therefore, the cardiac output is also decreased whether
the heart rate is normal or eventually increased. An essential
alteration in heart failure is the increase of peripheral resistance
which, together with the increased end-diastolic volume, is responsible for the marked increase in afterload. This mechanism - fundamental for the understanding of ventricular dynamics in heart failure - is the one to be interrupted in the therapy with vasodilators.

From the pathophysiologic mechanisms leading to cardiac failure,
mechanical overload is best understood[2]. It is due to pressure or
volume overload, as we see it frequently in hypertension, valvular
stenosis, valvular insufficiencies and shunts. Hypertrophy by addition of new myofibrils in parallel or in series are the common
compensatory mechanisms, but hypertrophy above a critical limit of
about 500g may be self destructive because the number of capillaries
do not grow in the same degree, leading to a O_2 diffusion limit
with consequent fibrosis. The myocardial reasons of heart failure
lie in a decreased quality of the muscle function mostly in conjunc-
tion with an increased percentage of interstitial fibrous tissue[3].

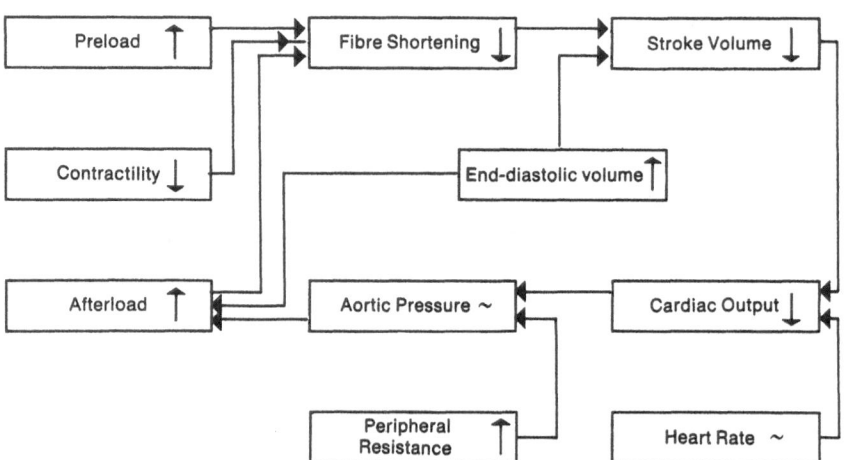

Fig. 1.   Constellation of physiopathologic parameters in congestive
          heart failure.

HOW TO PREVENT HEART FAILURE ?

Congestive heart failure is a lethal phenomenon. In the Framingham's study there was a 5-year mortality of 50% in patients with any class of heart failure. It is obvious that in advanced stages of heart failure mortality is even higher.

If we want to prevent heart failure, we have to view it in the context and as a matter of severity of heart disease. Preventing heart disease is therefore the first step in our effort to prevent heart failure. The following types of heart disease are partially apt to prevention:

- Hypertensive heart disease, which is in western civilization the most important precursor of heart failure[4], can certainly be prevented by early detection and adequate treatment of high blood pressure[5].
- Arteriosclerotic coronary heart disease can be diminished by lowering an elevated serum cholesterol. A decrease in the percentage of animal fat intake, abstention from smoking and regular exercises may have a preventive effect.
- Rheumatic heart disease. Streptocoque infections can be prevented or promptly treated in order to eliminate rheumatic myocarditis and rheumatic valvular heart disease. Good sanitary conditions are essential.
- Bacterial endocarditis can be prevented by early application of antibiotics.
- Pulmonary heart disease. Antipollution measures and a reduced body-weight seem valuable in order to prevent pulmonary heart disease.

HOW TO DELAY HEART FAILURE ?

Since heart disease can not always be prevented, our next step is to delay the onset of heart failure. This has to be done by adequate measures taken early in the stage of heart disease. Some rules of caution are:

- Regulating physical activities. If regular physical exercise may be a factor for preventing coronary heart disease, jogging or other strong exercises are contraindicated in our mind for patients with latent or overt heart failure.
- More consideration should be given to ingestion of salt. It is probable that the high incidence of hypertension is at least related in part to an unnecessary high salt intake, a habit in many societies around the world. A high salt intake certainly precipitates overt heart failure in patients with latent failure.
- Fluid balance should also be considered, especially intravenous sodium and fluid application has to be done with great caution

in patients showing early and also advanced stages of heart disease.
- Obesity has to be counteracted. Considerable loss of weight may
  delay or prevent entirely heart failure, for example in the
  Pickwick syndrome.
- High temperature especially coupled with high humidity may be the
  reason for massive decompensation of patients with latent heart
  failure. Also higher altitude (above 1500m) should be avoided in
  patients with severe heart disease.
- Cardiac arrhythmias, which often worsen heart failure, should be
  treated with the best available therapy.
- Careful consideration must be given to patients with heart disease
  who undergo a surgical procedure, specially with a long time of
  anesthesia. But heart surgery may itself be salutary if an important pressure or volume load can be taken off, eg. by resection of
  a ventricular aneurysm.
- Early drug therapy postpones the signs of decompensation. Drug
  therapy can be directed at an increase in contractility (digitalis),
  decrease of preload (diuretics, venodilators) or decrease of afterload with arterial vasodilators. Consumption of digitalis per
  inhabitant varies considerably among different countries, with a
  maximum in Germany. Since overdoses of digitalis are harmful,
  this drug should only be employed in patients with impaired myocardial function detectable by subtile signs of failure or in older
  patients who have to undergo important stresses as major surgery.
- Even diuretics diminish certain symptoms (dyspnea) and signs
  (oedema) of heart failure, they do not - except spironolactone -
  truly ameliorate the function of the heart and prolong often the
  circulation time[6], showing that they further diminish cardiac output.

Since in heart failure arterial and venous vasomotor tone is
considerably increased, vasodilator therapy may counteract the overshooting tone[7]. However, the development of tolerance of several
vasodilators is a serious drawback of most of these drugs. When
given to patients in e.g. class III and IV, it is at the moment impossible to say whether they prolong life. A careful individual
selection of the vasodilator and the dose is necessary.

SUMMARY

The answer to the question asked by Professor Kellermann is
simple: If we want to prevent heart failure, we have to try on
the basis of all our knowledge:

(1) to prevent heart disease, and if it is not possible
(2) to delay the onset of heart failure by hygienic measures, life
    style changes, drugs and eventually surgery.

We need, in accordance with Professor Kellermann's statement, a comprehensive heart care (C.H.C.) approach.

REFERENCES

1. W. Rutishauser and H. P. Krayenbühl, in:"Klinische Pathophysi-
    ologie," 4. Aufl., W. Siegenthaler, ed., Thieme Verlag,
    Stuttgart, 548 (1979).
2. S. Gunther and W. Grossmann, Determinants of ventricular function
    in pressure-overload hypertrophy in man, Circulation 59:679
    (1979).
3. O. M. Hess, J. Schneider, R. Koch, C. Bamert, J. Grimm and
    H. P. Krayenbühl, Diastolic function and myocardial structure
    in patients with myocardial hypertrophy, Circulation 63:360
    (1981).
4. W. B. Kannel, W. P. Castelli, P. M. McNamara, P. A. McKee and
    M. Feinleib, Role of blood pressure in the development of
    congestive heart failure, New Engl. J. Med. 287:781 (1972).
5. F. P. Epstein, Interventionsstudien der Hypertonie - eine
    Bestandesaufnahme bisheriger Ergenbnisse, Schweiz. Rundschau
    Med. 70:1396 (1981).
6. W. Rutishauser, F. Rhomberg and P. Sack, Über die Veranderungen
    der Kreislaufzeiten bei der Behandlung der haemodynamischen
    Herzinsuffizienz mit Diuretica und Herzglykosiden, Helv. Med.
    Acta 27:729 (1960).
7. W. Rustishauser, "The place of vasodilators in the long-term
    treatment of intractable heart failure," Huber, Bern (1981).

THE USE OF BETA-BLOCKERS IN SECONDARY PREVENTION

Lars Wilhelmsen, Anders Vendin, and Claes Wilhelmsson

Department of Medicine
Östra Hospital
S-416 85 Göteborg
Sweden

The basic problem in coronary heart disease (CHD) is a reduced myocardial blood supply relative to the myocardial energy demand. In acute myocardial infarction (MI) and sudden coronary death (SD) it is usually the blood supply that is reduced, but the demand might be enhanced under certain circumstances, as for example, during increased sympathetic activity (Wilhelmsen et al., 1978). Once the myocardium becomes ischemic, its metabolism switches utilization from fat to carbohydrate, which in turn, decreases energy production and reduces the contractile activity. The resulting reduction in mechanical performance may result in dilation of the heart and a reflex increase in general sympathetic tone and thus a further increase in the energy demand. The myocardium itself has a high content of noradrenaline, and ischemia is known to release these stores locally.

A direct result of the ischemic process, with hypoxia and ac-cumulation of noxious metabolites, is a change in myocardial cell membrane integrity, including among other things release of potassium, that rapidly leads to increased electrical excitability, with a shortening of the refractory period and a lowering of the fibrillation threshold. Relatively minor alterations in the balance between energy supply and demand (e.g., increased catecholamine release, arrhythmias) at the time of an acute coronary catastrophe can probably influence the ultimate size of the infarcted area. Thus, factors associated with the infarct size and the tendency for ventricular tachyarrhythmias to occur seem to be closely interrelated; clinical observations, for example, corroborate an association between infarct size and the occurrence of ventricular premature beats in myocardial infarction (Roberts et al., 1975).

241

Many studies have shown that the mortality during the first year after an MI is substantial. Thereafter it falls, but post-MI patients are still at a considerably higher risk than their healthy counterparts. It is also known, that post-MI patients are especially prone to die suddenly and most deaths are presumably caused by ventricular fibrillation (Vedin et al., 1975). In a study of a random population sample of men aged 47-54 years at entry and followed for 7 years, we found that 33% of the SD victims had suffered clinically verified MI before death.

Beta-blocking agents have been found to greatly reduce mortality when given prior to experimental coronary occlusion, but not when given afterwards (Pentecost and Austin, 1966; Khan et al., 1972; Kelliher et al., 1974). This may reflect the fact that beta-blockers are effective on the early (ischemic) arrhythmias, or in some way influence the ventricular fibrillation threshold, but they have so far not been particularly effective in reducing the more chronic arrhythmias. Some of these experimental findings and developing ideas provided a sufficient basis for trials of beta-blocking agents in survivors of MI.

Clinical trials of beta-blocking agents

Many studies with beta-blocking agents given after MI have been small, and by chance positive and negative results would be expected due to large confidence intervals, even if the above-mentioned hypothesis were true in man.

In an early study by Reynolds and Whitlock (1972), 87 patients were randomly given 400mg of alprenolol daily or placebo, and followed for one year. The study did not show any significant results.

In a study by Barber et al. (1975) treatment started immediately on admission to the hospital in 484 patients with a suspected or proven MI. A high dose of practolol was used (600mg daily) and the follow-up time was up to 24 months. There was no significant effect on mortality, but an analysis performed retrospectively pointed towards a possible effect in patients with entry heart rate above 100 beats/min.

In a study reported by Wilcox et al. (1980) three treatment groups were used, one treated with atenolol (n=127), one with propranolol (n=132), and one with placebo (n=128). The numbers were too small, the groups not fully comparable, and the follow-up time too short to show any effect on mortality.

In another trial comprising 783 patients randomized to propranolol and placebo after anterior MI, Baber (1979) did not either find any improvement of prognosis by beta-blockade.

A positive finding was first reported in a study of 230 patients, 57-67 years of age, randomly allocated to placebo or alprenolol and entering the trial between 1970 and 1971 in Göteborg, Sweden (Wilhelmsson et al., 1974). The number of sudden deaths was significantly reduced, but although the overall mortality was reduced by 50%, this reduction did not reach statistical significance. The pre-trial and within-trial exclusions totalled 20% of all these representative MI cases indicating that the drug can be used in 80% of MI cases in this age group.

Supportive evidence was presented from an open study by Ahlmark and Saetre (1976) also using alprenolol and two years of treatment.

The Multicentre International Study (1975, 1977) randomized 3.053 patients and the duration of treatment was up to two years, and the daily dose of practolol 400mg. The reduction of sudden deaths was significant, and reduction of total mortality was very close to significant (p=0.051). In addition, there was also a trend towards a reduction of the non-fatal reinfarction rate. As is well known the study had to be prematurely stopped because of side-effects.

A Danish study (Andersen et al., 1979) used alprenolol. All patients with a suspected or proven MI, regardless of age but stratified to below and above 65 years of age, were included in the one year study. As soon as the patient reached the emergency room of the hospital alprenolol or placebo was injected, and then oral alprenolol 400mg daily was given. After one year a significant reduction of mortality was shown in patients under the age of 65. No adverse effects were found in patients who did not develop MI. There was, however, a non-significant tendency towards a negative effect in patients above 65 years of age, among whom many had an unstable circulation.

During 1981 three different trials with beta-blockers were published. The Norwegian Multicentre Study Group (1981) found a strongly significant (p=0.0001) effect of timolol, 20mg daily, in 1.884 patients randomized to this drug (n=945) or placebo (n=939), with 98 and 152 deaths, respectively. They also demonstrated a significant decrease (p=0.0006) of the rate of non-fatal recurrencies of MI. The treatment was started 7-28 days after MI, and the patients were followed for 12-33 months.

The Beta-Blocker Heart Attack Trial in the USA (1981) randomized 3.837 patients after MI and followed 1.202 of them for 30 months. There was a mortality of 9.5% in the placebo group (183 deaths), and 7.0% in the propranolol group (135 deaths) and this difference is statistically significant (p=0.005).

The Gothenburg group (Hjalmarson et al., 1981) gave metoprolol, a beta$_1$ selective blocker, intravenously followed by oral adminis-

tration 100mg twice daily in a randomized study with 697 patients
with suspected acute MI on placebo and 698 on metoprolol. There
were 62 deaths in the placebo group (8.9%) and 40 deaths in the
metoprolol group (5.7%), a reduction of 36% (p=0.03) during the
three months of double-blind treatment, and found no adverse effects
in patients with suspected MI who did not develop an MI. This was
the first study to demonstrate a beneficial effect of beta-blockade
on survival during the early phase of MI. These patients are
followed for 2 years from the index MI in order to study whether an
early effect on infarct size determined with different methods might
also have an effect on long-term mortality.

Recently, Hansteen et al., (1982) have published results of a
double-blind study with propranolol 160mg daily in post-MI patients
with a fairly bad prognosis. Treatment started 4-6 days after the
infarction in 560 patients (278 on propranolol and 282 on placebo).
There was a significant effect on sudden death (11 in the propranolol
group and 23 in the placebo group (p=0.038), but there was no significant effect on total mortality (25 versus 37 deaths, p=0.12).

Comments

In conclusion, the studies presented support the concept that
beta-blocking drugs reduce short-term and long-term (up to two years)
mortality after MI. The exact mechanism of action is not defined.
Effects on myocardial metabolism probably both affecting the infarct
size and the tendency towards arrhythmias are plausible, but other
effects on for example thrombocyte aggregation can also be of some
importance. Beta-blockers with and without intrinsic stimulating
activity, both $beta_1$-selective and non-selective beta-blockers, and
those with and without membrane stabilizing properties have been
effective and due to the sizes of confidence limits for the various
studies there is no evidence for any of the drugs being better than
the other. Thus, the effect seems to be associated with the $beta_1$-
blocking property, and the choice of drug has to be based upon the
presence of this effect and lack of adverse side-effects of the risk
of such effects.

REFERENCES

Ahlmark, G., Saetre, H. 1976, Long-term treatment with beta-blockers
        after myocardial infarction. Eur. J. Clin. Pharmacol 10:77.
Andersen, M. P., Bechsgaard, P., Frederiksen, J., Hansenn, D. A.,
        Jürgensen, H. J., Nielsen, B., Pedersen, F., Pedersen-
        Bjergaard, O., Rasmussen, S. L. 1979, Effect of alprenolol on
        mortality among patients with definite or suspected acute
        myocardial infarction. Lancet II:865.

Baber, N. S. 1979, Multicentre propranolol post-infarction trial.
    Br. Heart J. 41:365.
Barber, J. M., Boyle, D. M., Chaturverdi, N. C., Singh, N., Walsh,
    M. J., 1975, Practolol in acute myocardial infarction. Acta.
    Med. Scand. (suppl) 587:213.
Beta-blocker Heart Attach Trial. JAMA, 6 November 1981, pp 2073-74.
    Coop. Trial. Prel. report. Nat. Heart, Lung, and Blood Inst.
Hansteen, V., Møinichen, E., Lorentsen, E., Andersen, A., Strøm, O.,
    Søiland, K., Dyrbekk, D., Refsum, A-M., Tromsdal, A.,
    Knudsen, K., Eika, C., Bakken, Jr. J., Smith, P., Hoff, P.I.
    1982, One year's treatment with propranolol after myocardial
    infarction: preliminary report of Norwegian multicentre
    trial. Br. Med. J. 284:155.
Hjalmarson, A., Elmfeldt, D., Herlitz, J., Holmberg, S., Málek, I.,
    Nyberg, G., Rydén, L., Swedberg, K., Vedin, A., Waagstein, F.,
    Waldenström, J., Wedel, H., Wilhelmsen, L., Wilhelmsson, C.
    1981, Effect on mortality of metoprolol in acute myocardial
    infarction. Lancet, p. 823, 17 Oct.
Kelliher, G. J., Widmer, C., Roberts, J. 1974, Seventh annual meeting
    of the international study group for research in cardiac
    metabolism. Abstract No 141, Quebec.
Khan, M. I., Hamilton, J. T., Manning, G. W., 1972, Protective
    effect of beta-adrenoceptor blockade in experimental coronary
    occlusion in conscious dogs. Am. J. Cardiol. 30:832.
A Multicentre International Study: 1975, Improvement in prognosis
    of myocardial infarction by long-term beta-adrenoceptor block-
    ade using practolol. Br. Med. J. II:735.
A Multicentre International Study: 1977, Reduction in mortality after
    myocardial infarction with long-term beta-adrenoceptor block-
    ade. Supplementary report. Br. Med. J. II:419.
The Norwegian Multicenter Study Group: 1981, Timolol-induced reduction
    in mortality and reinfarction in patients surviving acute
    myocardial infarction. N. Eng. J. Med. 304:801.
Pentecost, B. I., Austin, G. W., 1966, Beta-adrenergic blockade in
    experimental myocardial infarction. Am. Heart J. 72:790.
Reynolds, J. L., Whitlock, R. M. L. 1972, Effects of beta-adrenergic
    receptor blocker in myocardial infarction treated for one
    year from onset. Br. Heart J. 34:252.
Roberts, R., Husain, A., Ambos, H. D., Oliver, G. C., Cox, Jr. J. R.,
    Sobel, B. E. 1975, Relation between infarct size and ven-
    tricular arrhythmia. Br. Heart J. 37:1169.
Vedin, A., Wilhelmsson, C., Elmfeldt, D., Säve-Söderberg, J.,
    Tibblin, G., Wilhelmsen, L. 1975, Deaths and non-fatal re-
    infarctions during two years' follow-up after myocardial
    infarction. Acta. Med. Scand. 198:353.
Wilcox, R. J. Rowland, J. M., Banks, D. C., Hampton, J. R., Mitchell,
    J. R. A. 1980, Randomized trial comparing propranolol with
    atenolol in immediate treatment of suspected myocardial in-
    farction. Br. Med. J. 280:885.

Wilhelmsen, L., Vedin, A., Wilhelmsson, C. 1978, Beta blockade and
    sudden death following myocardial infarction. <u>Cardiovasc.</u>
    <u>Med</u>. 3:557.
Wilhelmsson, C., Vedin, A., Wilhelmsen, L., Tibblin, G., Werkö, L.
    1974, Reduction of sudden deaths after myocardial infarction
    by treatment with alprenolol. <u>Lancet</u> II:1157.

CLINICAL USE OF BETA-BLOCKERS AND CALCIUM-ENTRY

BLOCKERS IN STABLE ANGINA PECTORIS

Giuseppe Cocco*, Barbara Leishman,
Raffaele Pansini, and Carlo Strozzi

*Faculty of Internal Medicine
University of Basle, Switzerland
Department of Cardiology
General Medical Clinics and Therapeutics
University of Ferrara, Italy

SUMMARY

Drug therapy of coronary artery disease (CAD) is intended to eliminate the disproportion between oxygen supply and demand. Available drugs act mainly by reducing myocardial work and thus oxygen consumption. Some drugs antagonize the coronary spasm, thereby decreasing coronary resistance and, in consequence, increasing coronary blood flow.

Three classes of drugs have proved effective in the therapy of CAD: nitrates, β-blockers and calcium-entry blockers. The magnitude of oxygen consumption by the heart is determined by preload, afterload, heart rate and contractility. Nitrates, β-blockers and calcium-entry blockers reduce myocardial work while exerting fundamentally different effects on the cardiovascular system. Nitrates lower the preload, possess coronary dilating properties, can produce a more homogeneous pattern of coronary blood flow and, to a lesser extent, lower the afterload. β-blockers have a pronounced effect on the myocardium (reduce heart rate and contractility) and decrease blood pressure. They reduce coronary blood flow and do not antagonize coronary spasm. Calcium-entry blockers primarily lower the afterload, antagonize the coronary spasm and produce a more homogeneous coronary blood flow.

The treatment of CAD should be individualized: factors such type of symptoms, age, concomitant diseases and occupation influence the choice of the therapy.

In selected patients, nitrates, β-blockers and calcium-entry blockers can be combined, especially when there is arterial hypertension. Calcium-entry blockers are a heterogeneous class of drugs so that the choice of a given agent is not irrelevant. Nifedipine is more often combined with the β-blockers because it lacks electrophysiological effects. In contrast, verapamil may impair the sinus node function and prolongs the atrio-ventricular conduction. For this reason, there is a risk of precipitating bradycardia and/or heart failure by using verapamil together with a β-blocker. This combination is, therefore, rarely used.

INTRODUCTION

Coronary artery disease (CAD) generally results from anatomic coronary changes[1-3], but a functional component, so-called "coronary spasm"[4], may be an additional factor[5,6]. Our discussion here will deal mainly with typical and stable angina pectoris.

Irrespective of its cause, a narrowing of the coronary arteries reduces the coronary reserve and may produce a disproportion between myocardial blood flow and demand, thus inducing myocardial ischemia. This imbalance arises mainly when the oxygen demand is increased, eg. during exercise, emotional upset or exposure to cold. Less frequently, a coronary functional constriction, ie. "spasm"[4-6] may be the primary component triggering myocardial ischemia.

The available drugs act mainly by reducing myocardial work and thus oxygen consumption (Figure 1). Furthermore, some agents antagonize the coronary spasm, decrease coronary arterial resistance and can in consequence increase coronary blood flow.

Three classes of drugs have proved effective in the therapy of CAD: nitrates, β-blockers and calcium-entry blockers which are also called calcium antagonists, calcium-blockers or slow channel inhibitors.

The magnitude of cardiac oxygen consumption is determined by several parameters:

(1) left ventricular filling pressure and volume (preload);
(2) heart rate and myocardial contractility;
(3) peripheral resistance and systemic blood pressure.

Nitrates

Nitroglycerin has been in successful clinical use for nearly a century[7]. Although the good efficacy of the nitrates is indisputed[7-9], the precise mechanisms of their action in relieving myo-

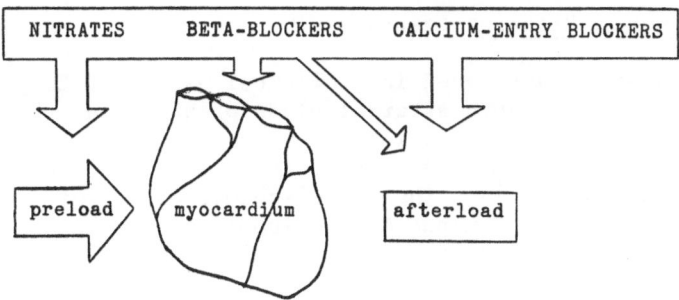

Fig. 1.   All three types of drugs reduce myocardial oxygen consumption, but their site of action differs, making it possible
to use these three agents in combination. Furthermore,
nitrates and calcium-entry blockers reduce coronary arterial resistance and are especially effective in patients
with myocardial ischemia caused by coronary spasm.

cardial ischemic pain in patients with CAD have been the subject of
considerable debate[8-9]. However, there is little doubt that the
anti-ischemic effect of the nitrates is predominantly due to their
action in reducing myocardial oxygen requirements. Nitrates primarily
lower the preload by dilating the veins (see Figure 1). In higher
doses, they can dilate the arterioles and reduce systemic arterial
resistance and thus the afterload. In patients with CAD, nitrates
may induce a myocardial blood flow redistribution in favor of the
subendocardium and the post-stenotic areas[8-10]. In such patients
although, global myocardial blood flow can decrease, regional differences in blood flow between post-stenotic and normal myocardial areas
diminish, since the decrease in blood flow is more marked in the normal than in the post-stenotic areas[8-10]. Therefore, nitrates produce
a more homogeneous pattern of myocardial blood flow, which effect
contributes to the anti-ischemic and thus pain-relieving properties
of the nitrates.

   In hearts with low filling pressure and impaired ventricular
function, the nitrate-induced decrease in cardiac filling can lead
to a fall in stroke volume and cardiac output, with possible hypotension and, in a few patients, syncope ("nitrate syncope").

   The effect of intravenous or sublingual nitroglycerin or isosorbidilate sets in rapidy (within 1-3 minutes) and lasts for up to
several hours. Oral (isosorbidilate) or percutaneous (nitroglycerin)
administration is used for a more prolonged effect, although to
acheive full therapeutic protection, oral nitrates must be given at
least q.i.d.

Beta-Adrenoreceptor Blocking Agents

β-blockers were introduced into clinical use about 20 years ago[11-14]. β-blockers reduce arterial blood pressure and thus the afterload. On acute administration, they do not reduce arterial resistance, but in long-term treatment arterial resistance decreases somewhat in most patients[15]. The anti-ischemic effect of β-blockers, however, is predominantly on the myocardium itself: they reduce heart rate and, to a lesser extent, myocardial contractility (see Figure 1). In coronary patients with left ventricular dyskinesia, β-blockers generally reduce dyskinetic motion and thus improve left ventricular function. In a minority of cases, however, they can induce left ventricular dysfunction and thus induce heart failure. With β-blockers a reduction in myocardial blood flow is usually observed. This effect seems to be related to vasoregulatory adjustments after decreased myocardial oxygen demands, rather than to vasoconstriction secondary to an unopposed alpha tone[10,15].

The introduction of the β-blockers was one of the major therapeutic advances of this century[12-15]. There can be no question about their efficacy in stable angina pectoris[12-15]. However, not all patients with CAD and angina pectoris benefit from or tolerate β-blockers. Furthermore, it has been theorized that β-blockers could actually enhance coronary spasm, at least in patients with vasospastic forms of myocardial ischemia[16]. Asthmatic patients with angina pectoris are seldom able to tolerate β-blockers: even cardioselective β-blockers like atenolol and acebutolol do not confer absolute safety[14,15] and there will be some asthmatic patients who present a worsening of their respiratory function. Similarly, coronary patients with functional or organic peripheral arterial disorders may experience exacerbation of the peripheral circulatory disorders during β-blocker treatment[14,15].

With regard to effectiveness, all β-blockers have proved valuable in the therapy of CAD. The choice of a particular β-blocker is therefore dictated by other reasons than efficacy[15]: cardioselectivity, lipophilicity (for the effect on the central nervous system), type of side effects and, not least in importance, pharmacokinetic behaviour.

Calcium-Entry Blockers

The term "calcium antagonists" was adopted on the basis of several pharmacologic observations, mainly becuase these substances inhibit some of the calcium-dependent effects on myocardial and smooth muscle cells[17-25]. Later the term "slow channel inhibitor" was proposed as a more accurate alternative[26]. However, both terms, although widely used, are scientifically incorrect as these agents do not actually antagonize the cellular effects of calcium[23,25-27].

Furthermore, there is no evidence that these agents inhibit slow
channels in tissues other than in the heart[28]. For these reasons,
the term "calcium-entry blockers" is better[28].

The process by which calcium-entry blockers inhibit calcium
movement across the cell membrane is not entirely understood[25-29].
All calcium-entry blockers have the common effect of reducing the
transmembrane transport of extracellular calcium ions on which vascular and myocardial tissues depend for contraction and impulse generation[19-29]. However, individually variable effects of the calcium-
entry blockers on smooth muscle, pacemaker tissue and myocardial
tissues make it impossible to group them together in a homogeneous
classification[19-29]. The marked differences among calcium-entry
blockers suggest either that the blocking action of each of these
drugs is relatively specific for different tissues, or that the
other actions of these agents account for their heterogeneous effects[19-29]. In Table 1 calcium-entry blockers are classified according to their chemical, pharmacological and electrophysiological
properties. For clinical purposes, the electrophysiological classification is probably the most important and useful.

Hemodynamic effects of calcium-entry blockers: calcium-entry
blockers dilate the arterioles and reduce systemic peripheral resistance, decrease end-systolic volume and pressure and, to a lesser
extent, end-diastolic volume and pressure. They thus reduce the
afterload (Figure 1). In contrast to nitrates, calcium-entry blockers
do not decrease (or only insignificantly) central venous pressure
and the filling pressure and therefore, they do not reduce the venous
reflux and the preload. The arterial vasodilation usually induces
an increase in stroke volume, generally accompanied by a modest increase in heart rate (slightly more pronounced with nifedipine and
nimodipine, negligible with therapeutic doses of verapamil and
diltiazem). At the same time, calcium-entry blockers decrease coronary arterial resistance and dilate the extramural coronary arteries,
thus increasing coronary blood flow. As shown with the Xenon technique, the calcium-entry blocker-induced coronary blood flow increase
is dependent on the baseline value, and these agents reduce the difference in flow between post-stenotic and normal areas, thus producing a more homogeneous myocardial blood flow[10,30]. On the other
hand, an intramural blood flow redistribution from the subendocardial
in favor of the subepicardial layers, which have a better perfusion,
cannot be excluded[10,31].

Electrophysiological effects of calcium-entry blockers[25,27,29]:
there are very marked differences between calcium-entry blockers in
terms of their electrophysiological effect on the myocardial tissues.
These differences serve as a basis for a clinical classification
(Table 1) and are a useful aid in selecting a calcium-entry blocker
for the therapy of patients with CAD, especially if calcium-entry
blockers are to be combined with other anti-anginal drugs.

Table 1. Classifications of Calcium-Entry Blockers[17-29,45-48]

Chemical Classification

1 aliphatic amines : verapamil, gallopamil, tiapamil, bencyclane, fendiline, prenylamine.

2 N-heterocycles : nifedipine, nimodipine, nitrendipine, diltiazem, suloctidil, perhexiline.

Pharmacologic Classification

1 diminishing only transmembrane calcium conductivity : nifedipine, nimodipine, nitrendipine.

2 acting additionally on channel kinetics : verapamil, gallopamil, tiapamil, diltiazem, fendiline.

Electrophysiologic Classification

1 negligible antiarrhythmic effect; no impairment of sinus node or atrioventricular function : nifedipine, nimodipine, nitrendipine.

2a effective against atrial tachy-arrhythmias; may impair sinus function; no effect on accessory atrioventricular pathways or ventricular pacemakers : verapamil, gallopamil, diltiazem, fendiline.

2b as 2a, effective also against some ventricular arrhythmias : tiapamil.

3 prolong the Q-T interval; may induce atypical ventricular tachycardia : prenylamine.

In brief, nifedipine is indicated mainly in patients with angina pectoris and impaired atrio-ventricular function, while verapamil, diltiazem and similar calcium-entry blockers have been found beneficial in coronary patients with atrial flutter of fibrillation, or in patients with supraventricular tachycardia associated with re-entry.

Therapeutic schedule: in clinical practice, nifedipine is used in the sublingual and oral form. The daily dosage of oral nifedipine is between 20-30 and 90-120mg, in three (sometimes two or four) equal doses. Verapamil, diltiazem, fendiline, prenylamine, perhexiline, tiapamil and gallopamil (some of these agents are not yet marketed) are not available for the sublingual therapy. Verapamil (and the two experimental drugs gallopamil and tiapamil) are also available for intravenous use, both for single dose and also for infusion, the latter being mainly restricted to the therapy of some arrhythmias and, rarely, of hypertensive crises or variant angina pectoris[24,29,32-35]. The dosage of verapamil is usually between 240 and 480mg/day per os, divided into 3-4 equal doses. That of diltiazem is between 225 and 450mg/day per os, usually divided in 3 equal doses. In the elderly, however, smaller doses of calcium-entry blockers will often be effective and induce fewer side effects.

Pharmacokinetics of calcium-entry blockers[19-25,27-29]: the most important data are summarized in Table 2. The bioavailability of verapamil, tiapamil and diltiazem is far from satisfactory: the therapeutic effect of these drugs will be less predictable than for drugs with a better bioavailability.

Possible Combinations

Nitrates and β-blockers have been successfully combined since the first years of their introduction: the combination has proved effective, well tolerated and advantageous in the majority of patients[13,15].

It should be stressed, however, that the dosage of nitrates and, to a certain extent, of β-blockers must be titrated individually. While some patients benefit from relatively high doses of nitrates and low doses of β-blockers, in other patients the opposite may be true. Furthermore, sublingual nitroglycerin has always been used to stop angina pectoris in patients treated with β-blockers, with success.

Nitrates and calcium-entry blockers can be successfully combined[19-25,36]. The two drugs act synergistically in dilating the coronary arteries and antagonize the coronary spasm. The nitrates act mainly by reducing the preload, while the calcium-entry blockers reduce the afterload. By using the two in combination, the net

Table 2. Pharmacokinetics of Some Calcium-Entry Blockers[19-25,27,29]

	Nifedipine	Verapamil	Tiapamil	Diltiazem
Daily dosage mg. p.o.	30-120	240-480	400-1,200	225-450
% absorption	>90	>90	>90	>90
% biovailab	65-70	10-25	10-25	10-20
Onset of effect				
i.v.	<5min.	<5min.	<5min.	unknown
sublingual	<5min.	unavailable	unavailable	unavailable
p.o.	<30min.	30-45min.	30-45min.	20-30min.
Therap. plasma concentr. ng/ml	7×10^{-8} 2×10^{-7}M	3.2×10^{-8} 2.0×10^{-7}M	unknown	7×10^{-8} 3×10^{-7}M
% protein binding	90	90	80	80
Plasma half-time				
α-phase, min.	150-180	15-30	15-30	15-25
β-phase, hr.	3-5	2-4.5	2-2.5	3-4
Metabolism, oral administration	extensive, to inert free acid and lactone	extensive, 12 metabolites by N-dealkylation and demethylation	extensive, similar to verapamil	extensive, desacetylated
Excretion, oral administration				
% renal	80	70	66	35
% fecal	15	15-20	30	65

Table 3.   Hemodynamic Effects of β-blockers and Calcium-Entry Blockers

	β-blockers	Calcium-entry blockers	β-blockers calcium-entry blockers
HR	D	u (i)	d
BP	D (d)	D (d)	D
SAR	I acutely u/d long-term	D	d (D)
SV	u (d)	I	I (i)
CO	D acutely d/u long term	I	I (i)
dp/dt	D (d)	i nifedipine u/d verapamil	u nifedipine d/D verapamil
EF	d acutely u long-term	I nifedipine i/u verapamil	u nifedipine d/D verapamil
LVSP	u (i)	D	D/d
LVEDP	i/u	u/d	u nifedipine i verapamil
PCWP	i/u	u/d	u nifedipine u/i verapamil
LVV	u/i	D	D nifedipine d/u verapamil
HRxBP	D	D	D
CBF	D	I	I/i

Abbreviations

HR = heart rate
SAR = systemic arterial resistance
dp/dt = dp/dt_{max}
LVSP = left ventricular systolic pressure
PCWP = pulmonary capilliary wedge pressure
HRxBP = pressure x rate product
d = slight decrease
I = significant increase
( ) = less common effect
long-term = results with long-term administration
verapamil = results with verapamil, gallopamil, tiapamil and diltiazem.

BP = blood pressure
SV = stroke volume
CO = cardiac output
EF = ejection fraction
LVEDP = left ventricular end-diastolic pressure

LVV = left ventricular volume
CBF = coronary blood flow
D = significant decrease
u = unchanged
i = slight increase
acutely = results with acute administration
nifedipine = results with nifedipine

N.B. All data were obtained with therapeutic sublingual (nifedipine) and oral (all drugs) doses (our data and references 19-25,29).

effect is enhanced while lower doses of each are required, thereby
reducing the frequency and severity of side effects and thus improving patient compliance.

Several authors have successfully combined β-blockers with
nifedipine[19-25,26]. The synergistic and counterregulatory effects
of this combination are summarized in Table 3. There is no doubt
that, in many patients, the combination of a β-blocker and nifedipine
has a great oxygen sparing effect, more so than with the single
drugs.

Of course nitrates can be added, and thus one can also decrease
the preload.

On the other hand, some patients do not tolerate these combinations as they lower blood pressure excessively, signs of heart
failure may appear and angina pectoris may be increased in intensity
and frequency[37]. There are several reasons for the exacerbation of
angina pectoris: either the combination of β-blockers and nifedipine
excessively lowers the pre-stenotic coronary perfusion pressure
thereby decreasing the post-stenotic flow and intensifying myocardial
ischemia or, in predisposed patients with poor ventricular function,
this combination might further attenuate cardiac function producing
an increase in left ventricular volume and wall tension whereas the
opposite effect would be appropriate. These side effects are, however, rare, in our experience having occurred in only 4/127 patients
treated with this combination.

Verapamil, diltiazem and tiapamil prolong the atrio-ventricular
impulse conduction and, especially in patients with sick sinus syndrome, may impair the sinus node function. The β-blockers induce
similar electrophysiological effects. Therefore, calcium-entry
blockers of this type should not be combined with the β-blockers,
particularly not with those with marked electrophysiological effects,
eg. propranolol, nadolol and sotalol. With such a combination there
is a definite risk of precipitating a major bradycardia[38,39]. It
should be added, however, that this side effect is more probable
when verapamil is given intravenously to a patient treated with a
β-blocker and with an acute myocardial infarction[38]. Severe bradycardia is much rarer when verapamil is given orally to a coronary
patient already receiving treatment with a modern β-blocker (eg.
atenolol). In selected patients with CAD and supraventricular tachyarrhythmias however, we deliberately use a combination of atenolol
and verapamil, with good efficacy (suppression of the arrhythmia)
and without untoward reactions. In such patients verapamil is to
be preferred to nifedipine, because of their different antiarrhythmic properties.

It has been observed that, in rare cases, the combination of
nifedipine and β-blockers may precipitate or aggravate left ven-

tricular failure[37]. In our experience, the risk of hypotension
with the combination of nifedipine and a β-blocker is greater in
patients who are also being treated with nitrates or/and diuretics.
This is not surprising: the net effect of these drugs used together
is to reduce preload, afterload, heart rate and myocardial contractility, thereby favouring the development of hypotension and heart
failure.

In therapeutic doses, neither nifedipine[19-25] nor verapamil[24,25]
exerts any marked negative effect. The effect may be different
however if the calcium-entry blockers are given to coronary patients
treated with β-blockers.

Recent Studies

In recent years, we have been assessing the effect of verapamil
(5-10mg i.v., or 120mg t.i.d. per os) on the ejection fraction of
coronary patients. The ejection fraction was assessed by means of
^{99m}-Tc radionuclide angiography. In those doses, the effect of
verapamil on the ejection fraction was usually insignificant. Later,
in the same patients receiving long-term therapy with a β-blocker
(oxprenolol, sotalol, atenolol or pindolol), the β-blocker did not
induce any major changes in the ejection fraction. At this point
verapamil was combined in a dose of 5mg intravenously or 80mg t.i.d.
per os. A variable effect was observed: where the baseline ejection
fraction (under β-blockade) was 40% or more, verapamil had no effect
on left ventricular contractility (42 patients). However, in those
cases with a baseline ejection fraction between 30 and 39%, verapamil
caused the ejection fraction to decrease by more than 15% in 9/35
patients, and in three of these nine patients, signs of cardiac
failure were detected (gallop rhythm or signs of pulmonary congestion). Verapamil was stopped and, after a washout period, nifedipine
(10mg sublingually or 10mg t.i.d.) was combined to the β-blocker.
The ejection fraction decreased by more than 15% in only 1/35
patients and no signs of cardiac failure were detected. It appears
therefore that the risk of a negative inotropic effect in a coronary
patient treated with a β-blocker is greater with verapamil than
nifedipine, at least at the above doses. Thus, calcium-entry blockers to be given to a patient with CAD being treated with a β-blocker
must be selected carefully with due consideration to both the electrophysiological and the hemodynamic effects of both drugs in the
combination. It goes without saying that such a combination is
contraindicated in hypotensive patients or in those with poor cardiac
function. Finally, the possible role of other drugs, eg. antiarrhythmic or tricyclic agents, must also be taken into consideration.

CALCIUM-ENTRY BLOCKERS AND β-BLOCKERS IN STRABLE ANGINA PECTORIS:
COMPARATIVE TRIALS

Recent controlled trials have compared the therapeutic value
of calcium-entry blockers and β-blockers in stable angina pectoris.
Dragie et al.[40] have reported the results of a comparison between
nifedipine and propranolol, Frishman et al.[41] a comparison between
verapamil and propranolol, Cocco et al.[42] a comparison between
nifedipine and pindolol, and De Ponti et al.[43] a comparison of
verapamil, nifedipine and propranolol. Despite the different in-
vestigational approaches used, the results were very similar: both
the calcium-entry blockers and the β-blockers were shown to be useful
anti-anginal drugs, and the calcium-entry blockers appear to be a
safe and effective alternative to β-blockers in patients with stable
angina pectoris. There are pronounced differences between the
calcium-entry blockers and the β-blockers: the anti-anginal effect
of the β-blockers appears earlier[42], and the calcium-entry blockers
are superior in decreasing resting myocardial ischemia[42]. Similar
results have been reported with the calcium-entry blocker lido-
flazine[44].

SOME REFLECTIONS ABOUT ANTI-ANGINAL DRUGS

It is an established principle that the treatment of the coronary patient should be individualized[22-24]. Factors such as type
of symptom, age, concomitant diseases, occupation and physical activity influence the choice of therapy. Nitrates, β-blockers and
calcium-entry blockers are the mainstays of the modern drug therapy
of CAD. Of course, preventive health care, individualized physical
training, psychotherapy and, in selected cases, anti-coagulants and
platelet aggregation inhibitors, help the patient both to understand
his illness and to augment his exercise tolerance while reducing
myocardial ischemia.

Most cardiologists use nitrates with a β-blocker to treat stable
angina pectoris occurring on effort. However, in angina in which
coronary spasm is believed to play a major role, many cardiologists
would prefer to use a calcium-entry blocker sooner, often together
with nitrates. However, not all patients with angina pectoris
benefit from or tolerate the β-blockers or the calcium-entry blockers.

The presence of exertional cardiac arrhythmias accompanying
angina pectoris would favour β-blockers as drugs of first choice.
Where β-blockers are contraindicated, however, eg. in asthma, a
calcium-entry blocker is to be preferred. Patients with severe
angina pectoris, especially in the presence of hypertension are
those who will obtain the greatest benefit from the combined use
of β-blockers and calcium-entry blockers. In hyperkinetic syndromes,
a β-blocker and verapamil or diltiazem are theoretically preferable
to nifedipine. In hypokinetic situations, nifedipine will probably
be superior.

Lastly, in the presence of severe angina pectoris, with left main coronary obstruction, three vessel obstruction or obstruction of two vessels including the proximal left anterior descending coronary artery, surgery should be considered unless significant additional surgical risks are identified.

ACKNOWLEDGEMENT

The authors thank Dr. Marcel Eckert, F. Hoffmann-La Roche & Co. Ltd., Basle, for checking the pharmacokinetic data.

REFERENCES

1. Z. Vlodaver, K. Amplatz, H. W. Burchell and J. E. Edwards, "Clinical, Angiographic & Pathologic Profiles. Coronary Heart Disease," Springer-Verlag, New York, Heidelberg, Berlin (1976).

2. F. F. Battle and C. A. Bertolasi, eds., "Cardiopatie isquémica," 3rd Edition, Inter-Médica, Buenos Aires (1980).

3. W. Ganz, Editorial: coronary spasm in myocardial infarction: fact or fiction?, Circulation 63:487-488 (1981).

4. R. N. MacAlpin, Correlation of the location of coronary arterial spasm with the lead distribution of ST segment elevation during variant angina, Amer. Heart J. 99:555-564 (1980).

5. A. Maseri, S. Severi, M. De Nes, A. L'Abbate, S. Chierchia, M. Marzilli, A. M. Ballestra, O. Parodi, A. Biagini and A. Distante, "Variant" angina: one aspect of a continuous spectrum of vasospastic myocardial ischemia. Pathogenic mechanisms, estimated incidence and clinical and coronary arteriographic findings in 138 patiens, Amer. J. Cardiol. 42:1019-1035 (1978).

6. A. Maseri, A. Malliani, R. Paoletti, A. L'Abbate, P. Bobba, S. Chierchia, O. Parodi and B. Marino, Ruolo del vasospasmo nella cardiobatia ischemia, Boll. Soc. Ital. Cardiol., Suppl. Simposii, Roma, 23-26 maggio, pp. 229-293 (1981).

7. J. B. Stetson, Intravenous nitroglycerin: a review, Internat. Anesthes. Clin. 16:261-298 (1978).

8. R. C. Klein, T. M. Grehl, K. B. Stengert and D. T. Mason, Evaluation of the effects of systemic nitroglycerin or perfusion of ischemic myocardium in coronary heart disease assessed intraoperatively by antegrade blood flow through intact saphenous vein bypass grafts, Amer. Heart J. 101:292-299 (1981).

9. R. L. Feldman and R. C. Conti, Editorial: relief of myocardial ischemia with nitroglycerin: what is the mechanism?, Circulation 64:1098-1100 (1981).

10. H. J. Engel and P. R. Lichtlen, Beneficial enhancement of coronary blood flow by nifedipine. Comparison with nitroglycerin and beta blocking agents, Amer. J. Med. 71:658-666 (1981).

11. J. W. Black and J. S. Stephenson, Pharmacology of a new adren-
    ergic β-receptor blocking compound (nethalide), Lancet 2:311-
    314 (1962).
12. R. P. Ahlquist, Adrenergic beta-blocking agents, Fortschr.
    Arzneimittelforsch 20:27-43 (1976).
13. D. F. Weetman, A review of the actions and clinical uses of
    β-adrenoceptor blocking drugs, Drugs of Today 12:261-305
    (1977).
14. G. Cocco, C. Strozzi and D. Chu, I farmaci beta-bloccanti sono
    stati introdotti in clinica 15 anni or sono: miti e realtà,
    Cl. Ter. 80:67-89 (1977).
15. J. M. Cruickshank, The clinical importance of cardioselectivity
    and lipophilicity in beta blockers, Amer. Heart J. 100:160-178
    (1980).
16. H. R. Hellstrom, Evidence in favor of the vasospastic cause of
    coronary artery thrombosis, Amer. Heart J. 97:449-452 (1979).
17. A. Fleckenstein, On the basic pharmacological mechanism of
    nifedipine and its relation to therapeutic efficacy, in:"3rd
    International Adalat Symposium," A. D. Jatene and
    P.R. Lichtlen, eds., Excerpta Medica, Amsterdam-Oxford, pp.1-
    13 (1976).
18. M. Kohlardt and A. Fleckenstein, Inhibition of the slow inward
    current and tension in mammalian ventricular myocardium,
    Naunyn-Schmiedeberg's Arch. Pharm. 298:267-272 (1977).
19. 1st International Nifedipine "Adalat" Symposium, K. Hashimoto,
    E. Kimura and T Kobayashi, eds., Univ. of Tokyo Press (1973).
20. 2nd Internationa Adalat Symposium, W. Lochner, W. Braasch and
    G. Kroneberg, eds., Springer-Verlag, Berlin, Heidelberg, New
    York (1975).
21. 3rd International Adalat Symposium, New Therapy of Ischemic
    Heart Disease, A. D. Jatene and P. R. Lichtlen, eds., Excerpta
    Medica, Amsterdam-Oxford (1976).
22. 4th International Adalat Symposium. New Therapy of Ischemic
    Heart Disease, P. Puech and R. Krebs, eds., Excerpta Medica,
    Amsterdam-Oxford-Princeton (1980).
23. Calcium antagoinists (Ca-Blockers). Pharmacological, Physio-
    logical and Clinical Aspects, B. W. Johansson, J. Abelin and
    G. Flygt, eds., Acta Pharmacol Toxicol 43 Suppl. 1 (1978).
24. I Calcio-Antagonisti in Cardiologia, Atti della Soc. Ital.
    Cardiol, Boll. Soc. Ital. Cardiol. Suppl. Segreteria della
    Società Roma, ed., 39 Congresso, Milano 30 settembre - 3
    ottobre (1978).
25. Symposium on Nifedipine and Cacium Flux Inhibition in the
    Treatment of Coronary Arterial Spasm and Myocardial Ischemia,
    B. Lown, ed., Amer. J. Cardiol. 44:779-844 (1979).
26. W. G. Nayler, Cacium antagonists, Eur. Heart J. 1:225-237 (1980).
27. Calcium-entry blockers in coronary artery disease, Circulation
    65 Suppl. I, 1-59 (1982).
28. P. M. Vanhoutte, Calcium-entry blockers and vascular smooth
    muscle, Circulation 65 Suppl. I, 11-19 (1982).

29. 1st International Symposium on Tiapamil, B. N. Singh, B. Leishman
    G. Cocco and G. Haeusler, eds., Cardiology, Suppl. April 8-10,
    1981 (in press).
30. H. J. Engel and P. R. Lichtlen, Effect of nifedipine on regional
    myocardial blood flow on coronary patients at rest and during
    rapid atrial pacing, pp. 55-63 in Ref. 21.
31. H. J. Engel, Discussion, pp. 76-77 in Ref. 21.
32. M. K. Heng, B. N. Singh, A. H. G. Roche, R. M. Norris and
    C. J. Mercer, Effects of intravenous verapamil on cardiac
    arrhythmias and on the electrocardiogram, Amer. Heart J. 90:
    487-498 (1975).
33. B. N. Singh, J. T. Collett and C. Y. C. Chew, New perspectives
    in the pharmacologic therapy of cardiac arrhythmias, Progr.
    Cardiovasc. Dis. 22:243-301 (1980).
34. G. Cocco, D. Chu and C. Strozzi, Dimeditiapramin (Ro 11-1781)
    a new calcium antagonist, in the management of supraventricu-
    lar tachyarrhythmias in patients with acute myocardial in-
    farction, Clin. Cardiol. 2:131-134 (1979).
35. T. Menzel, P. Kirchner, G. Cocco and D. F. Gasser, Die parente-
    rale Behandlung von supraventrikulären und ventrikulären
    Herzrhythmusstörungen mit dem Kalziumantagonisten Tiapamil,
    Z. Kardiol. 70:163-171 (1981).
36. P. Schlup, Behandlung mit Depot-Nitraten und Calcium-Antagonis-
    ten, Schweiz med. Wschr. 111:449-454 (1981).
37. Side effects and general discussion, pp. 284-291 in Ref. 21.
38. F. Hagemeijer, Verapamil in the management of supraventricular
    tachyarrhythmias occurring after a recent myocardial infarc-
    tion, Circulation 57:751-755 (1978).
39. B. Subramanian, M. J. Bowles, A. B. Davies and E. B. Raftery,
    Combined therapy with verapamil and propranolol in chronic
    stable angina, Amer. J. Cardiol. 49:125-132 (1982).
40. H. J. Dargie, P. G. Lynch, D. M. Krikler, L. Harris and
    S. Krikler, Nifedipine and propranolol: a beneficial drug
    interaction, Amer. J. Med. 71:676-682 (1981).
41. W. H. Frishman, N. A. Klein, J. A. Strom, H. Willens,
    T. H. LeJemtel, J. Jentzer, L. Siegel, P. Klein, N. Kirschen,
    R. Silverman, S. Pollack, R. Doyle, E. Kirsten and
    E. H. Sonnenblick, Superiority of verapamil to propranolol
    in stable angina pectoris: a double-blind randomized cross-
    over trial, Circulation 65:51-59, Suppl. I (1982).
42. G. Cocco, C. Strozzi, D. Chu, R. Amrein and E. Castagnoli, The
    therapeutic effects of pindolol and nifedipine in patients
    with stable angina pectoris and asymptomatic resting ischemia,
    Eur. J. Cardiol. 10:59-69 (1979).
43. C. De Ponti, F. Mauri, G. R. Ciliberto and B. Caru, Comparative
    effects of nifedipine, verapamil, isosorbide dinitrate and
    propranolol on exercise-induced angina pectoris, Br. J. Clin.
    Practice Symposium Suppl. 8 (1980).
44. F. M. Howard and D. T. Mallegol, The management of stable angina
    pectoris with lidoflazine - a new calcium antagonist: a

double-blind placebo-controlled clinical evaluation, Curr.
Ther. Res. 30:181-193 (1981).

45. P. F. Fazzini, F. Marchi and P. Pucci, Ventricular tachycardia:
a new iatrogenic possibility, Amer. Heart J. 90:805-806 (1975).

46. R. Puritz, M. A. Henderson, S. N. Baker and D. A. Chamberlain,
Ventricular arrhythmias caused by prenylamine, Br. Med. J.
2:608-609 (1977).

47. B. Aloisi, A. Stuto, R. Negro, E. Mossuti, M. Moncada and
B. Brancati, Contributo casistico alla conoscenza della
tachicardia ventricolare a torsione di punte, Boll. Soc. Ital.
Cardiol. 23:345-361 (1978).

48. N. Riccioni, C. Bartolomei and S. Soldani, Prenylamine-induced
ventricular arrhythmias and syncopal attacks with Q-T prolongation, Cardiology 66:199-203 (1980).

PLATELET FUNCTION AND COAGULATION IN SECONDARY PREVENTION OF

ISCHEMIC HEART DISEASE

A. Strano and G. Davi

Institute of Clinical Medicine
University of Palermo
Italy

The possibility of secondary prevention of myocardial infarction
using drugs as anticoagulants and platelet antiaggregants attracted
much attention in the last few years. Results obtained with anti-
coagulants until now, however, did not always correspond to expec-
tations and were quite contradictory although recent studies show that
such treatment is quite effective[1] and also that the overall incidence
of brain injuries following prolonged anticoagulant treatment does
not increase: in fact a certain increase of brain haemorrhages is
counterbalanced by a decrease of thrombotic events[2].

Studies on platelet antiaggregants are quite numerous but results
are still contradictory. AMIS[3] failed to show a decrease of coronary
mortality and of sudden death due to coronary injury after aspirin
administration. A certain improvement was instead shown by PARIS[4],
particularly after administration of an aspirin-dipyridamole association. The American trial with sulphinpyrazone presented an evident
decrease of mortality, particularly during the first months of treatment[5]. The recent study by Polli[6] showed that sulphinpyrazone is
quite effective in reducing overall reinfarction incidence.

This brief review of recent trials with anticoagulant and antiaggregating agents suggests some considerations on the reasons why
such studies were performed. We may in fact ask ourselves whether
patients with previous myocardial infarction are really characterized
by coagulative or platelet activation to such a degree as to justify
treatment with anticoagulant or antiaggregating agents.

The majority of evidence supports the hypothesis that myocardial
infarction is caused by thrombotic occlusion of a diseased coronary
artery proximal to the site of infarction[7,8]; thus one clinical

263

complication that could be influenced by drugs that inhibit thrombus
formation is the formation of occlusive thrombi and myocardial infarcts.
Our present knowledge is incomplete concerning the mechanisms that
initiate different types of thrombosis and the dominant reaction involved; exposed collagen, released ADP, thromboxane A$_2$, epinephrine
and thrombin may all take part in thrombus formation, but their
relative importance in each situation has not been assessed. Drugs
that affect one type of thrombosis may have little effect on another
type of thrombosis.

It is generally recognized that when the endothelium is damaged,
platelets interact with the subendothelial constituents and that it
is this interaction that sets the stage for thrombus growth in arteries.
Platelet interaction with collagen leads to the release of ADP and
activation of the arachidonate pathway with the formation of thromboxane A$_2$ [9]. Both ADP and thromboxane A$_2$ cause circulating platelets to change shape and adhere to each other and to the platelets
already attached to the damaged wall. The growth of a thrombus
appears to depend on the generation of thrombin at the injury site
and around the aggregating platelet mass that has adhered to the sub-
endothelium. Thrombin causes the release of platelet granule contents
and activates the arachidonate pathway[10]; in addition thrombin causes
aggregation and release through a mechanism that is independent of
ADP release or the formation of thromboxane A$_2$ [11]. Thus thrombin
causes further platelet aggregation and release. Another agent that
is important in thrombus formation is the ADP that is released from
platelets or lost from damaged cells at the injury site or from red
blood cells that may hemolyse. Born[12] recently presented evidence
that red blood cells hemolysis may play a part in the initiation of
thrombi under some experimental conditions. Studies in experimental
animals have shown that epinephrine is another agent that may play a
part in the formation of intravascular platelet aggregates and
thrombi[13].

Available data on coagulation phenomena in patients with previous
myocardial infarction are not numerous. It was recently reported that
patients with previous myocardial infarction show a remarkable increase
of plasma levels of fibrinopeptide A which is quite a sensitive marker
for coagulative activation; such increase is even more remarkable than
that found in patients with angina[14]. In an interesting study by
Stormorken[15] coronarography performed on apparently healthy subjects
showed that antithrombin III levels were lower in subjects with A blood
ary vascular lesions and that angiopositive subjects with A blood
group presented lower values of antithrombin III. Angiopositive
patients show an increase of factor VIII antigen and levels are significantly different in subjects with O blood group as compared to
those with A blood group[15].

Studies on possible platelet activation in patients with previous
myocardial infarction are quite numerous and only the most significant

ones will be considered in the present review, including some data obtained by our own study group.

It should be pointed out that investigation of platelet activation involves consideration of various phenomena since platelet functions are manifold and we cannot avail ourselves of a single technique for the detection of several parameters of platelet activation. Individual phenomena linked with platelet aggregation, however, may be studied.

An increase of aggregability with various inducers[16] has already been shown in patients with ischemic heart disease during exercise but such increase may be effectively counteracted by pretreating patients with aspirin. In our study on patients with acute myocardial infarction[17] we showed that spontaneous platelet aggregation occurs in these patients while it never occurs in normal subjects.

Detection of circulating platelet aggregates using the technique devised by Wu and Hoak allows further studies on platelet aggregation. Circulating platelet aggregates[18] are increased particularly in patients with unstable angina and myocardial infarction; such increase can also be detected in the first few days after myocardial infarction[19]. The value of this technique is confirmed by the fact if platelet aggregates are assayed in patients with spontaneous angina[20] an increased concentration of such aggregates will be found only during ischemic attacks and only in the coronary sinus, but not in peripheral blood.

Our knowledge of platelet aggregability may be further improved by determining platelet sensitivity to prostacyclin. In our study on patients with acute myocardial infarction[21] platelet sensitivity to prostacyclin was found to be decreased in the first few days (Figure 1) after acute infarction; 50% inhibition of platelet aggregation with ADP may be achieved with about 3 ng prostacyclin, a remarkably higher amount than that found in control subjects (about 1ng). Sensitivity to prostacyclin (PGI_2) and to other prostaglandins (PGE_1 and PGD_2) is decreased also in patients with angina and returns to basal values only 1 hour after angina attacks[22].

Evaluation of platelet release corresponding to an increased release of specific factors from platelets is also possible. Platelet factor 4 and betathromboglobulin, which are platelet-specific alpha-granule proteins, have been shown to be readily released from platelets and their appearance in plasma can be sensitively detected by radioimmunoassay[23,24]. Moreover these proteins may reflect the release of the platelet derived growth factor, which is also contained in the alpha-granule[25] and which is postulated to be the mediating mitogen in at least some types of arterial lesion formation.

An increase of betathromboglobulin (BTG) was found in patients with ischemic heart disease and particularly in those with myocardial

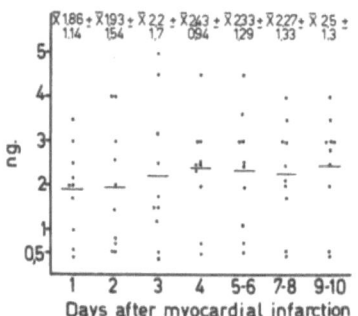

Fig. 1.   Prostacyclin needed for 50% platelet aggregation inhibition.

infarction[26]. In our study on patients with acute myocardial
infarction BTG was found to be increased[27] during the whole period
of observation. No significant increase of BTG was usually found in
normal subjects during exercise[28] but there was an evident increase
of BTG in patients with ischemic heart disease and positive effort
test; such augment is less marked in patients with ischemic heart
disease but with negative effort test.

        Platelet factor 4 (PF4), also released by platelets, was found
in our study to be increased during the first few days following
acute infarction[17]. The presence of further risk factors, such as
diabetes, should also be considered, since this condition involves
a significant increase of plasma levels of PF4, as has already been
shown in a previous report[29]. High plasma levels of platelet factor
4 were found[30] in patients with spontaneous angina undergoing cather-
ism; such levels were found to decrease after 1 month treatment with
nitrates and propranolol.

        Further important information on platelet breakdown may be ob-
tained by determining platelet survival. Abnormal platelet consumption in patients with vascular disease reflects platelet accumulation
at sites of damaged vascular endothelium, a circumstance possibly
predisposing to downstream microembolization. Shortened in vivo platelet survival in about 50% of patients with angiographically documented
coronary atherosclerosis and stable angina has been shown[31,32,33,34].
Fuster also reported[33] shortened platelet survival times in asymptomatic

cigarette smokers with a positive family history of coronary artery
disease. Evidence that this platelet consumption occurred in the
coronary vasculature rather than the vasculature at large, however,
remains cirumstantial because platelet survival studies do not distinguish between coronary artery disease and antherosclerosis
generally.

Platelet survival was found to be much shorter in patients with
myocardial infarction in whom coronarography had shown normal arteriographic features[35]; this finding could suggest a direct participation
of platelets in the occlusive process.

Platelets play also an important role in the metabolism of arachidonic acid, eventually yielding thromboxane A_2, a potent vasocon-
stricting and platelet aggregating agent, through cyclic AMP modula-
tion. Prostacyclin synthesis occurs at vessel walls starting from
arachidonic acid, and its effects are opposite to those of thromboxane
A_2. Arachidonic acid present in the platelet membrane is converted to
cyclic endoperoxides by action of enzyme cyclo-oxygenase; during the
early phases of aggregation these cyclic endoperoxides are converted
to thromboxane A_2 by action of enzyme thromboxane synthetase[36,37].
Thromboxane A_2 is highly labile and can be measured only with bioassay;
its stable metabolite thromboxane B_2 can be measured with radioimmunoassay and other techniques.

The following experiment by Smith[38] suggests that thromboxane A_2
synthesis may play an important role in the pathogenesis of some acute
coronary events, such as sudden coronary death : arachidonic acid administration to rabbits is followed by sudden death within 2 or 3
minutes; a marked increase of plasma levels of 6-keto PGF1alpha (a
prostacyclin metabolite) and of thromboxane B_2 (stable thromboxane
metabolite) was then found. Simultaneous administration of a selective inhibitor of thromboxane synthetase causes a remarkable prolong-
ment of survival time in these animals.

In patients with past myocardial infarction thromboxane A_2 synthesis is often increased[39]. In our experience patients with acute
myocardial infarction often show a progressive increase of thromboxane formation from platelets[21]; such increase is more marked than
in control subjects, on the 7th or 8th day after acute infarction
(Figure 2). We already hinted at the importance of the so-called
risk factors such as hyperlipidemia (particularly hypercolesterolemia)
which causes a further increase of serum thromboxane synthesis, that
attains higher levels than in type IV and in controls[40].

Our attention was also focused on plasma thromboxane levels which
were always higher in patients with acute myocardial infarction than
in healthy normal subjects[21]. During pacing[41] in patients with effort
angina, plasma thromboxane levels increased, while such increase was
not evident in plasma from control subjects. Another important

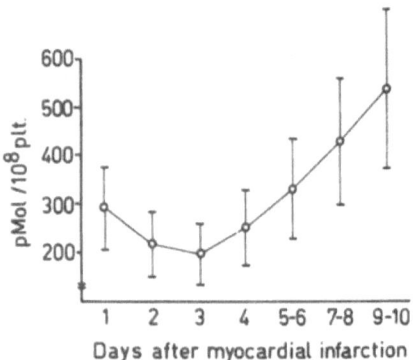

Fig. 2.   T x B$_2$ formation by platelets after thrombin stimulation.

parameter can be obtained from simultaneous determination of plasma
thromboxane in samples drawn from coronary sinus and aorta[42].   Such
ratio was found to be increased only in patients with spontaneous
angina a few hours after angina attacks and not in control subjects,
in patients with non-ischemic heart disease or in patients with
angina who had suffered pain more than 24 hours earlier.

Plasma prostacyclin determinations should finally be considered
although data from literature are still quite contradictory. Decreased
plasma prostacyclin levels were found in patients with myocardial infarction[43]; plasma 6-keto PGF$_{1alpha}$ levels may also be determined but
a comparative evaluation of concentrations in coronary sinus and aorta
failed to show any difference between patients and controls[42].   Moreover prostacyclin administration in patients with ischemic heart
disease[44] causes a remarkable decrease of platelet aggregability but
fails to reduce the number of spontaneous attacks in patients with
spontaneous angina if compared to placebo administration.

In conclusion these data suggest therefore the occurrence of
coagulative and platelet activation in patients with previous myocardial infarction. Whether a primary or secondary event, platelet
reactivity in the atherosclerotic and narrowed coronary vessels could
lead to limitation of blood flow and platelet thrombus formation and
subsequent development of myocardial ischemia; products of myocardial
ischemia could lead to further platelet activation and a vicious cycle
would start. Increased production of thromboxane by platelets may
be an important factor in vasococtriction, platelet agregation and
vascular damage[45]. Excess of thromboxane has been implicated as a

potential factor in the genesis of vasospasm; reduced synthesis of prostacyclin by atherosclerotic plaques may be pathogenetic in vaso-constriction related to unchecked actions of thromboxane.

At the same time, however, caution is necessary in evaluating the relative importance of these factors, since they are certainly not the only ones that may influence the prognosis of patients with myocardial infarction; treatment with anticoagulant or antiaggregating agents, although well scheduled, will not be sufficient in these cases.

REFERENCES

1. Report of the Sixty Plus Reinfarction Study Research Group, A double blind trial to assess long-term oral anticoagulant therapy in elderly patients after myocardial infarction. Lancet 2 : 989,(1980).
2. Report of the Sixty Reinfarction Study Research Group, Risks of long term oral anticoagulant therapy in elderly patients after myocardial infarction, Lancet, 1: 64,(1980).
3. Aspirin Myocardial Infarction Study Research Group, A randomized controlled trial of aspirin in persons recovered from myocardial infarction, J.A.M.A., 243: 661,(1980).
4. The Persantine Aspirin Reinfarction Study Research Group, Persantine and aspirin in coronary heart disease, Circulation, 62: 449, (1980).
5. The Anturane Reinfarction Trial Research Group, Sulphinpyrazone in the prevention of cardiac death after myocardial infarction, N. Engl. J. Med., 298: 289, (1978).
6. Anturan Reinfarction Italian Study, Sulphinpyrazone in post-myocardial infarction, Lancet, 1: 237, (1982).
7. A.B. Chandler, I. Chapman, L.R. Erhrardt, W.C. Roberts, C.J. Schwartz, D. Sinapius, D.M. Spain, S.Sherry, P.M. Ness, T.L. Simon, Coronary thrombosis in myocardial infarction. Report of a workshop on the role of coronary thrombosis in the pathogenesis of acute myocardial infarction, Am. J. Cardiol., 34: 823, (1974).
8. W.F.M. Fulton, D.J. Sumner, Causal role of coronary artery thrombotic occlusion in myocardial infarction : evidence of stereo-arteriography, serial sections and ^{125}I fibrinogen autoradiography, Am. J. Cardiol., 39: 322, (1977).
9. M.Hamberg, J. Svensson, B. Samuelsson, Thromboxanes: a new group of biologically active compounds derived from prostaglandin endoperoxides, Proc. Nat. Acad. Sci. USA, 72: 2994, (1975).
10. H.Holmsen, Prostaglandin endoperoxide-thromboxane synthesis and dense granule secretion as positive feedback loops in the propagation of platelet responses during "the basic platelet reaction", Thromb. Haemostas, 38: 1030, (1977).

11. J. P. Cazenave, J. F. Mustard, Mechanisms of platelet shape change, aggregation and release induced by collagen, thrombin or A23187, J. Lab. Clin. Med., 90: 707,(1977).

12. G. V. R. Born, A. Wehmeier, Inhibition of platelet thrombus formation by chlorpromazine acting to diminish hemolysis, Nature, 282:212, (1979).

13. J. I. Haft, P. D. Kranz, F. J. Albert, K. Fani, Intravascular platelet aggregation in the heart induced by norepinephrine. Microscopic studies, Circulation, 46: 698, (1972).

14. G. G. Neri Serneri, G. F. Gensini, R. Abbate, C. Mugnaini, S. Favilla, C. Brunelli, S. Chierchia, O. Parodi, Increased fibrinopeptide A formation and thromboxane A2 production in patients with ischemic heart disease: relationship to coronary pathoanatomy, risk factors and clinical manifestations. Am. Heart J., 101: 185, (1981).

15. H. Stormorken, The thrombo-haemorrhagic balance. Acta Med. Scnad (Suppl.), 642: 131, (1980).

16. J. Mehta, P. Mehta, C. J. Pepine, C. R. Conti, Platelet function studies in coronary artery disease. VII. Effect of aspirin and tachycardia stess on aortic and coronary venous blood, Am. J. Cardiol., 45: 945, (1980).

17. G. Davi, M. Traina, S. Novo, V. Albano, G. L. Piraino, M. P. Muzzo, G. Marano, A. Raineri, A. Strano, Platelet function changes in acute myocardial infarction, Proc. II Eur. Symp. on Coagulation, Platelet Function, Fibrinolysis and Vascular Disease. Palermo December 5-8, in press. (1980).

18. M. B. Schwartz, J. Hawiger, S. Timmons, G. C. Friesinger, Platelet aggregates in ischemic heart disease, Thromb. Haemostas., 43: 185, (1980).

19. P. Mehta, J. Mehta, Platelet function studies in coronary artery disease: evidence for enhanced platelet microthrombus formation activity in acute myocardial infarction, Am. J. Cardiol. 43: 757, (1979).

20. R. M. Robertson, D. Robertson, G. C. Friesinger, S. Timmons, J. Hawinger, Platelet aggregates and coronary sinus blood in patients with spontaneous coronary artery spasm, Lancet, 2: 829, (1980).

21. A. Strano, G. Davi, M. Traina, S. Novo, A. Raineri, Thromboxane formation by platelets and platelet sensitivity to prostacyclin in patients with acute myocardial infarction, Thromb. Haemostas, 46: 759, (1981).

22. H. Sinzinger, G. Schernthaner, J. Kaliman, Sensitivity of platelets to prostaglandins in coronary heart disease and angina pectoris, Prostaglandins, 22: 773, (1981).

23. A. E. Bolton, C. A. Ludlam, D. S. Petter, J. Moore, J. D. Cash, A radioimmunoassay for platelet factor 4, Thromb. Res., 8: 51 (1976).

24. K. L. Kaplan, H. L. Nossel, M. Drillings, G. Lesznik, Radioimmunoassay of platelet factor 4 and Betathromboglobulin: development and application to studies of platelet release reaction in

relation to fibrinopeptide A generation, Br. J. Haematol., 39: 129, (1978).

25. D. R. Kaplan, F. C. Chao, C. D. Stiles, H. N. Antoniades, C. D. Scher, Platelet alpha granules contain a growth factor for fibroblasts, Blood, 53: 1043, (1979).

26. T. C. Smitherman, M. Milam, J. Woo, J. T. Willerson, E. P. Frenkel, Elevated betathromboglobulin in peripheral venous blood of patients with acute myocardial ischemia: direct evidence for enhanced platelet reactivity in vivo, Am. J. Cardiol., 48: 395, (1981).

27. G. Davi, M. Traina, S. Novo, A. Pinto, G. L. Piraino, G. Marano, R. Alfano, Formazione di malondialdeide piastrinica e livelli plasmatici di betatromboglobulina in pazienti con infarto del miocardio acuto, Giorn It. Cardio., XI : 1238, (1981).

28. J. Mehta, P. Mehta, Comparison of platelet function during exercise in normal subjects and coronary artery disease patients: potential role of platelet activation in myocardial ischemia, Am. Heart J., 103: 49, (1982).

29. G. Davi, G. B. Rini, M. Averna, S. Novo, G. Di Fede, A. Mattina, A. Notarbartolo, A. Strano, Enhanced platelet release reaction in insulin-dependent and in insulin-independent diabetic patients, Haemostasis, in press, (1982).

30. S. P. Levine, J. Lindenfield, J. B. Ellis, N. M. Raymond, L. S. Krentz, Increased plasma concentrations of platelet factor 4 in coronary artery disease, Circulation, 64: 626, (1981).

31. P. Steele, J. Rainwater, R. Vogel, E. Genton, Platelet-suppressant therapy in patients with coronary artery disease, J.A.M.A., 240: 228, (1978).

32. J. L. Ritchie, L. A. Harker, Platelet and fibrinogen survival in coronary atherosclerosis. Response to medical and surgical therapy, Am. J. Cardiol., 39: 595, (1977).

33. J. H. Chesebro, D. Fuster, R. L. Frye, Smoking, family history and shortened platelet survival in coronary disease patients age 50 and under, Circulation, 58, suppl. II: 221, (1978).

34. D. J. Doyle, C. N. Chesterman, J. F. Cade, J. R. McGready, G. L. Rennie, F. J. Morgan, Plasma concentrations of platelet specific proteins with platelet survival, Blood, 55: 82, (1980).

35. P. Steele, J. Rainwater, R. Vogel, Abnormal platelet survival time in men with myocardial infarction and normal coronary arteriogram, Am. J. Cardiol., 41: 60, (1978).

36. P. Needleman, G. Kaley, Cardiac and coronary prostaglandin synthesis and function, N. Engl. J. Med., 298: 1122, (1978).

37. S. Moncada, J. R. Vane, Arachidonic acid metabolites and the interactions between platelets and blood vessels, N. Engl. J. Med., 300: 1142, (1979).

38. J. B. Smith, H. Araki, A. M. Lefer, Thromboxane A$_2$,Prostacyclin and aspirin: effects on vascular tone and platelet aggregation, Circulation, 62, Suppl. V: 19.

39. A. Szczeklik, R. J. Gryglewski, J. Musial, L. Grodzinska, M. Serwonska, E. Marcinkiewicz, Thromboxane generation and plate-

let aggregation in survivals of myocardial infarction, Thromb. Haemostas, 40: 66, (1978).

40. A. Strano, G. Davi, M. Averna, G. B. Rini, S. Novo, G. Di Fede, A. Mattina, A. Notarbartolo, Platelet sensitivity to prostacyclin and thromboxane production in hyperlipidemic patients. Thromb. Haemostas, 48: 1, (1982).

41. M. Tada, T. Kuzuya, M. Inoue, K. Kodama, M. Mishima, M. Yamada, M. Inui, H. Abe, Elevation of thromboxane B_2 levels in patients with classic and variant angina pectoris, Circulation, 64: 1107, (1981).

42. P. D. Hirsh, L. D. Hillis, W. B. Campbell, B. G. Firth, J. T. Willerson, Release of prostaglandins and thromboxane into the coronary circulation in patients with ischemic heart disease, N. Engl. J. Med., 304: 685, (1981).

43. G. G. Neri Serneri, Aterosclerosi e stati trombofilici, Pozzi ed., Roma, (1979).

44. S. Cherchia, C. Patrono, E. Crea, G. Ciabattoni, R. De Caterina, G. A. Cinotti, A. Distante, A. Maseri, Effects of intravenous prostacyclin in variant angina, Circulation, 65: 470, (1982).

45. S. D. Gertz, G. Uretsky, R. S. Wajnberg, N. Navot, M. S. Gotsman, Endothelial cell damage and thrombus formation after partial arterial constriction: relevance to the role of coronary arter spasm in the pathogenesis of myocardial infarction, Circulation, 63: 476. (1980).